Welcome Note

Whispers of Nature: 100 Poems on Climate and Environment is a poetic journey through the often-overlooked beauty of our planet. This collection celebrates the delicate balance and majesty of nature, capturing its essence with urgency and reverence. Through verses inspired by autumn's colors, resilient blooms, and the serene dance of clouds, the poems highlight the critical need for environmental preservation. Each page invites readers to connect with nature's spirit and serves as a call to action to protect and cherish our ecosystems. This book is both a tribute to nature's splendor and a plea for environmental stewardship, aiming to inspire a commitment to preserving our planet's precious wonders.

Mushila Victor Isaacs

Cover designed by the Author.

© Victor Mushila, 2024
Nairobi – Kenya.

All rights reserved.
No parts of this book may be reproduced or stored in any retrieval system, or transmitted in any form or by any means; electronically, mechanically, by photocopying, or otherwise without the prior written permission of the author or, on his behalf.

Thank you for respecting the author's
intellectual property rights.

Introduction

Welcome to *Whispers of Nature: 100 Poems on Climate and Environment*, a collection that invites you to embark on a journey through the profound and often overlooked beauty of our natural world. Each poem within these pages is a testament to the delicate balance and vibrant majesty of our planet, capturing its essence with both reverence and urgency.

Here, you will find inspiration in the fleeting brilliance of autumn's colors, the resilient bloom of flowers in harsh winters, and the serene dance of clouds across the sky. From the majestic cliffs that guard ancient songs to the tranquil beauty of rainforests and the profound interconnectedness of life, these poems celebrate the wonder of our environment and underscore the critical need for its preservation.

As you turn each page, you will be invited to connect deeply with the spirit of nature, from the enchanting allure of the seaside to the hidden magic of valleys and the silent whispers of trees. These verses are not just a reflection of nature's splendor but a call to action—urging us to protect, cherish, and nurture the delicate ecosystems that sustain life on Earth.

Whispers of Nature* is more than a collection of poems; it is a heartfelt plea for environmental stewardship and a beacon of hope for a harmonious future. Let these words inspire you to embrace the beauty of our planet and join in the vital work of preserving its precious wonders for generations to come.

M.V.I.

Dedication

To the Almighty God, the Creator of all things, whose magnificent handiwork is revealed in every corner of our world. "For since the creation of the world God's invisible qualities—His eternal power and divine nature—have been clearly seen, being understood from what has been made." - Romans 1:20.

To the Earth, our shared home, whose boundless beauty and resilience inspire every word in this collection.

To the United Nations Environment Programme (UNEP), whose unwavering commitment to safeguarding our planet and promoting sustainable practices has been a beacon of hope in a world that urgently needs it. Your efforts to preserve and protect our environment are a testament to the power of collective action and the profound impact of dedicated stewardship.

To the gentle whisper of the wind, the steadfast strength of the mountains, and the quiet wisdom of the ancient trees—may your voices continue to echo through the hearts of all who seek to listen.

To the future generations, may you inherit a world still rich in the wonders of nature, a world where every creature, great and small, thrives in harmony.

And to those who tirelessly protect and nurture our planet, your dedication is the hope that sustains us all. This book is for you.

Acknowledgements

I am deeply grateful to all those who have contributed to the creation and completion of *Whispers of Nature: 100 Poems on Climate and Environment*. My heartfelt thanks go to my family and friends for their unwavering support and encouragement throughout this journey. Your belief in this project has been a source of strength and inspiration.

A special thanks to the dedicated environmentalists and activists whose work continues to shed light on the urgent need for environmental stewardship. Your tireless efforts in preserving our planet have profoundly influenced this collection.

I also wish to acknowledge the work of the United Nations Environment Programme (UNEP) for their invaluable contributions to raising awareness about environmental issues and promoting sustainable practices. Your commitment to safeguarding our planet has inspired many, including myself.

My gratitude extends to the editors and reviewers who helped refine and polish these poems, ensuring that they convey the beauty and urgency of our natural world. Your insights and expertise have been invaluable.

Lastly, to every reader who picks up this book, thank you for joining me on this journey through the wonders and challenges of our environment. May these poems inspire you to embrace and protect the beauty that surrounds us.

With sincere appreciation,
M.V.I.

Content

1. A Tapestry of Stars 9
2. Anthem of the Atmosphere 10
3. Autumn's Embrace 11
4. Beyond the Horizon 12
5. Blossoms in the Snow 13
6. Breath of the Biosphere 14
7. Canvas of Clouds 15
8. Chorus of the Canopy 16
9. Chorus of the Cliffs 17
10. Cradle of Creation 18
11. Earth Echo 19
12. Earth's Dreams 20
13. Earth's Harmony 21
14. Earth's Heartbeat 22
15. Earth's Voice 23
16. Echoes of the Earth 24
17. Embrace of the Ecosystem 25
18. Gleam of the Grasslands 26
19. Harmony of Habitats 27
20. Harmony of the Hills 28
21. Harmony of the Horizon 29
22. Hues of the Horizon 30
23. Lament of the Lakes and Rivers 31
24. Lullaby of the Leaves 32
25. Mountain Majesty 33
26. Murmur of the Mountains 34
27. Ode to the Ozone 35
28. Ode to the Symphony of the Seas 36
29. Pulse of the Planet 37
30. Radiance of Rainforests 38
31. Reflections in the Rain 39
32. Reflections in the River 40
33. Reflections of the Rain 41
34. Rivers of Hope 42
35. Seaside Serenade 43
36. Seasons of Change 44
37. Serenade of the Shoreline 45
38. Shadows of the Shore 46
39. Silent Echoes 47
40. Song of the Savanna 48
41. Song of the Snowcaps 49
42. Symphony of Nature 50

43.	Symphony of Rain	51
44.	Symphony of Soil	52
45.	Symphony of the Seas	53
46.	Symphony of the Seasons	54
47.	Tears of the Earth	55
48.	The Blooming Biome	56
49.	The Blossoming Bog	57
50.	The Bountiful Basin	58
51.	The Celestial Canopy	59
52.	The Drowsy Deltas	60
53.	The Dying Desert	61
54.	The Earth Song	62
55.	The Earth Weeps	63
56.	The Echoing Forest	64
57.	The Edge of Dawn	65
58.	The Eternal Ecosystem	66
59.	The Fragile Forest	67
60.	The Glowing Grove	68
61.	The Graceful Grassland	69
62.	The Green Canopy	70
63.	The Hidden Hills	71
64.	The Infinite Isles	72
65.	The Last Leaf	73
66.	The Last River	74
67.	The Last Standing Tree	75
68.	The Melting Mountains	76
69.	The Peaceful Peaks	77
70.	The Placid Plains	78
71.	The Quiet Quarry	79
72.	The Radiant Reef	80
73.	The Rising Ridge	81
74.	The Roaring River	82
75.	The Rustling Reeds	83
76.	The Sacred Sands	84
77.	The Sacred Skies	85
78.	The Secret Stream	86
79.	The Serene Savanna	87
80.	The Silent Indian Ocean	88
81.	The Silent Sky	89
82.	The Silent Springs	90
83.	The Silent Stream	91
84.	The Singing Sand	92
85.	The Singing Summit	93

86.	The Throbbing Thicket	94
87.	The Thundering Thaw	95
88.	The Tranquil Tropics	96
89.	The Veiled Valley	97
90.	The Verdant Valley	98
91.	The Wailing Waves	99
92.	The Waterfall's Whisper	100
93.	The Weeping Wetlands	101
94.	The Whispering Wild	102
95.	Whisper of the Trees	103
96.	Whispering Willows	104
97.	Whispers in the Woods	105
98.	Whispers of the Plains	106
99.	Whispers of the Wind	107
100.	Whispers of Wilderness	108
101.	Contact the Author	111

1. A Tapestry of Stars

Beneath the sky, a tapestry of stars,
We weave our dreams, near and far.
Each twinkle tells a tale of old,
Of earth and sky, of warmth and cold.

Mountains rise with ancient grace,
Their peaks a testament to time and space.
Forests whisper secrets deep,
In their shadows, creatures sleep.

Rivers carve their winding path,
Through valleys green, they gently laugh.
Oceans vast with waves that roar,
Guard the secrets of the ocean floor.

Deserts stretch with golden sands,
A silent dance across the lands.
In every grain, a story lies,
Of sunlit days and starry skies.

The air we breathe, so pure and clear,
A gift from nature, held so dear.
In every breath, a life renewed,
A promise kept, a future viewed.

Stars above, a guiding light,
Through darkest days and endless night.
They remind us of our place,
In this vast and wondrous space.

So let us cherish earth and sky,
Protect the land, the sea, the high.
For in this tapestry of stars,
Lies the future, ours and ours.

This poem is meant to inspire a sense of wonder for the night sky and the importance of protecting our view of the stars against the encroaching light pollution.

2. Anthem of the Atmosphere

Anthem of the Atmosphere, a breath above,
Cradling the earth with winds of love.
A veil of blue, so vast, so wide,
In its embrace, the world does hide.

The sky sings out, in hues of morn,
In the Atmosphere's anthem, life is born.
A canvas for the sun's fiery trail,
For the moon's soft glow, the stars' tale.

This anthem plays, in whispers of air,
In the rustle of leaves, in the eagle's stare.
A symphony of climate, of weather's weave,
In the Atmosphere's song, we believe.

Yet, it weeps for the smoke that mars its face,
For the toxins we emit, a disgrace.
The Anthem of the Atmosphere, a cry for change,
For cleaner skies, within our range.

Let us join in chorus, with actions true,
To sing the anthem, to renew.
For in the air we breathe, the sky we see,
Lies the key to life, for you and me.

So let's uphold this anthem, with care, with might,
For the Atmosphere's song, is our guiding light.
A pledge to protect, to honor, to revere,
The Anthem of the Atmosphere.

This poem is written to serve as a call to action to respect and protect the atmosphere, the very air we breathe and the sky that shelters all life on Earth.

3. Autumn's Embrace

When summer's warmth retreats with grace,
And leaves begin their downward chase,
The world is touched by a painter's brush,
In hues of amber, a tranquil hush.

Autumn's embrace is soft and kind,
A tapestry of colors intertwined.
Crimson, gold, and pumpkin spice,
Each leaf a stroke, each chill a slice.

The air is crisp, the sky is clear,
Nature's whisper, we can hear.
It tells of change, of cycles spun,
Of time's passage, of seasons done.

Beneath the boughs, the ground is dressed,
In a quilt of leaves, we find our rest.
The world slows down, takes a gentle pace,
In the loving arms of autumn's embrace.

So let us walk through this painted scene,
Where every step stirs a memory keen.
And hold this moment, soft and brief,
In the falling glory of the autumn leaf.

This poem captures the essence of autumn's fleeting beauty, celebrating its vibrant colors and ephemeral grace. It inspires us to cherish each moment of this enchanting season and embrace the profound lessons found in its transitory splendor.

4. Beyond the Horizon

Where the sky kisses the sea, in a line so fine,
Lies a realm unseen, beyond the horizon's spine.
A place of wonder, where dreams intertwine,
With the threads of the future, in a design divine.

Beyond the horizon, where the unknown does recline,
Adventures await, in a world without confine.
It calls to the brave, to those who incline,
To discover what lies past that distant line.

The sun sets there, in a spectacle that does shine,
Promising a tomorrow, fresh and benign.
It's the edge of possibility, a border so prime,
Where hope takes flight, beyond the horizon's climb.

So let us journey, with hearts aligned,
To the vast beyond, where life is redefined.
For at the edge of sight, where our dreams are signed,
New horizons await, in the vastness, unconfined.

This poem is to inspire thoughts of the endless possibilities and adventures that wait just out of sight, encouraging a look beyond the familiar.

5. Blossoms in the Snow

Amidst the winter's frosty glow,
Where icy breezes tend to blow,
A sight so rare, a vibrant show,
The delicate blossoms in the snow.

Petals soft of pink and white,
Against the snow, a contrast bright.
They stand in defiance of the cold's might,
A testament to nature's fight.

Each bloom a spark of life's embrace,
A touch of spring's forthcoming grace.
In the chill, they find their place,
And add to winter's icy lace.

So let us cherish this tender sight,
Of blossoms braving the wintry night.
For even in the cold's harsh throw,
There's beauty, like blossoms in the snow.

This poem captures the delicate beauty and unwavering resilience of flowers that defy the harshest winters to bloom, symbolizing hope and strength in the face of adversity. It inspires us to find grace and courage in even the most challenging circumstances.

6. Breath of the Biosphere

In the heart of the forest, where whispers reside,
The breath of the biosphere flows far and wide.
Leaves like lungs, in a rhythmic dance,
Inhale the sun, in nature's trance.

From the depths of the ocean, to the tallest tree,
A symphony of life, wild and free.
Phytoplankton bloom in the ocean's embrace,
While forests stand tall, in a verdant grace.

The wind carries whispers of ancient lore,
Of ecosystems thriving, from mountain to shore.
Each breath a promise, each sigh a song,
In the balance of nature, where we all belong.

Carbon captured, oxygen released,
A cycle of life, that never ceased.
In the warmth of summer, in winter's chill,
The breath of the biosphere, steady and still.

So let us cherish this delicate thread,
Of life interconnected, by which we're fed.
For in the breath of the biosphere, we find,
The essence of life, beautifully intertwined.

This poem inspires a profound appreciation for the delicate balance and breathtaking beauty of our planet's biosphere, reminding us to cherish and protect the intricate web of life that sustains us all.

7. Canvas of Clouds

Upon the sky, a canvas wide,
Where clouds in whispers softly glide.
Brushstrokes of white on azure blue,
Nature's art, a timeless view.

In dawn's embrace, they blush with light,
A palette warm, a gentle sight.
As day unfolds, they drift and play,
In patterns grand, they weave the day.

Storms may gather, dark and bold,
A tempest tale, a story told.
Yet even in their fiercest form,
They bring the rain, life to transform.

At twilight's hush, they softly fade,
In hues of pink, a serenade.
The stars emerge, the night takes hold,
The canvas dark, with dreams untold.

So let us gaze and find our peace,
In clouds that float, in skies that cease.
For in their dance, we see the ties,
Of earth and air, of seas and skies.

This poem inspires a deep connection with the ever-changing artistry of clouds, celebrating their graceful dance across the sky and their vital role in sustaining life on Earth. It reminds us to appreciate the beauty and purpose woven into nature's canvas above us.

8. Chorus of the Canopy

In the heart of the forest, where the sunlight dapples,
A symphony rises, in rustles and apples.
The leaves whisper secrets, in hushed, verdant tones,
A chorus of the canopy, in breaths and moans.

Majestic trees stand, with arms stretched to the skies,
Their leaves like a choir, where the wind harmonizes.
Each branch, a verse of an ancient, rooted song,
Singing of the earth, where they've thrived so long.

Beneath the green arches, life teems with grace,
A delicate balance, in this sacred space.
From the tiniest insect to the birds that soar,
Each life plays a note in the forest's core.

The chorus of the canopy, a testament to time,
A melody of nature, pure and sublime.
It sings of a world, untouched and free,
A call to preserve, for eternity.

So let us listen, and heed the call,
To protect these havens, for one and for all.
For the chorus of the canopy, in its infinite sweep,
Is the earth's loving lullaby, that puts our hearts to sleep.

This poem resonates with the profound beauty and vital importance of our natural environments, inspiring us to cherish and protect the precious landscapes that sustain and nourish life.

9. Chorus of the Cliffs

Upon the edge where earth meets sky,
The cliffs stand tall, as gulls fly by,
A chorus of stone, silent and strong,
In their presence, we belong.

They've watched the ages come and go,
Felt the sun's warmth and winter's snow,
A testament to time's grand flow,
The cliffs sing songs of high and low.

Their faces carved by wind and wave,
A symphony of the brave,
Where eagles nest and spirits crave,
The chorus of the cliffs, we save.

But now they face a threat unseen,
From actions harsh, from hands unclean,
We must protect these sentinels old,
For in their chorus, our tale is told.

So let us sing with cliffs so grand,
For they are part of our land,
In their chorus, a story cast,
Of a future built to last.

This poem stirs a deep sense of awe for the majestic cliffs, urging us to preserve their ancient song and honor the timeless beauty they embody. It inspires a commitment to safeguarding these towering guardians of nature for generations to come.

10. Cradle of Creation

In the dawn of time, where life began,
A cradle of creation, nature's grand plan.
Mountains rose, and rivers carved,
A world of wonders, beautifully starved.

From fertile valleys to ocean's deep,
Life emerged from its ancient sleep.
Forests whispered, deserts sang,
In the cradle of creation, where life first sprang.

The sun's warm kiss, the moon's soft glow,
Nurtured the seeds that began to grow.
In the dance of seasons, life took flight,
In the cradle of creation, day and night.

Storms may rage, and fires may burn,
Yet life persists, with each return.
In the heart of nature, a resilient beat,
In the cradle of creation, life is sweet.

So let us honor this sacred ground,
Where the breath of life is always found.
For in the cradle of creation, we see,
The essence of life, wild and free.

This poem awakens a deep sense of responsibility and awe for the delicate, profound beginnings of our natural world. It inspires us to protect and nurture the fragile foundations that sustain the beauty and wonder of life on Earth.

11. Earth Echo

In the silence of the world, when the clamor fades,
The Earth Echo resounds, through the glades and glens.
A reverberation of life, a pulse, a beat,
A sound that travels, with every feat.

The echo of the rain, as it kisses the soil,
The murmur of the trees, in their stoic toil.
The hum of the bees, in their tireless quest,
The Earth Echo speaks, and we are the guest.

It's the rhythm of the waves, crashing on the shore,
The whisper of the wind, through the canyon's roar.
The call of the wild, the cry of the free,
The Earth Echo is the voice, of the land and the sea.

But this echo is fading, growing dim with time,
Muffled by the noise, of our industrial chime.
We must listen closely, and heed its plea,
To protect the echo, and let it be.

For the Earth Echo is our heritage, our past,
A reminder of what's at stake, what must last.
It's the song of our planet, a hymn so profound,
In the echo of the Earth, our future is bound.

So let's raise our voices, in a chorus so bold,
To amplify the echo, a thousandfold.
For in the Earth Echo, we find our kin,
A symphony of life, that we're all in.

This poem emphasizes the interconnectedness of all life and inspires a commitment to protecting the natural world, reminding us that every living being plays a crucial role in the symphony of existence.

12. Earth's Dreams

In the quiet of the night, when the world is still,
The Earth dreams in colors, with a painter's skill.
A dream of green forests, and deep blue seas,
Of clear skies above, and the hum of bees.

The Earth dreams of rivers, flowing free and strong,
Of mountains standing proud, where they belong.
It dreams of the deserts, with their shifting sands,
And the icy poles, in their frozen lands.

In its dreams, there's balance, a perfect harmony,
Where every creature thrives, wild and free.
The Earth dreams of a future, where we all care,
For the soil, the water, and the precious air.

But the Earth also dreams of the scars we've made,
The forests we've felled, the price we've paid.
It dreams of the smoke, the plastic, the waste,
Of the beauty we've lost, in our reckless haste.

Yet, in its wisdom, the Earth still dreams of hope,
Of a world united, where we can cope.
With the challenges we face, together as one,
To heal the planet, under the same sun.

So let's join the Earth, in its dreaming quest,
To restore its health, to give our best.
For the Earth's dreams are ours to fulfill,
A call to action, a test of our will.

This poem inspires a profound sense of responsibility and optimism, urging us to create a sustainable future that honors and fulfills the Earth's deepest dreams. It calls us to align our actions with the planet's aspirations, nurturing a harmonious and enduring world for generations to come.

13. Earth's Harmony

In the cradle of cosmos, our planet spins,
A sphere of blue and green, where life begins.
A symphony of ecosystems, in delicate balance,
Earth's harmony, a dance of chance and dalliance.

From the peaks where eagles dare to soar,
To the ocean depths, with mysteries galore.
The forests breathe, the deserts blaze,
All tuned to Earth's harmonious ways.

The rustling leaves, the roaring seas,
The buzzing bees, the whispering breeze.
Each element in concert, with nature's score,
A melody that thrives from core to shore.

But this harmony is fragile, under threat,
By our own hands, a looming debt.
We must act, with care and empathy,
To preserve this tune, for eternity.

Let's embrace sustainability, with open arms,
And safeguard our planet from harm.
For Earth's harmony is our greatest treasure,
A gift of life, beyond measure.

So let us unite, in conservation's song,
And heal the world, to which we belong.
For in Earth's harmony, we find our place,
A symphony of survival, for the human race.

This poem inspires a profound sense of unity and responsibility, calling us to protect and preserve the delicate balance of our planet's ecosystems. It encourages us to work together in safeguarding the intricate web of life that sustains us all.

14. Earth's Heartbeat

In the stillness of the wild, where the world's pulse is felt,
Throbs the Earth's heartbeat, where all life is dealt.
A rhythm so profound, in the soil and the air,
A beat that resounds, from here to everywhere.

The Earth's heartbeat is steady, in the roots that entwine,
In the flutter of wings, in the forest's pine.
It's the drum of the rain, on the leafy canopy,
The cadence of the seasons, in their endless symphony.

The heartbeat is strong, in the ocean's waves,
In the whisper of the grass, in the caves.
It's the vibration of life, the energy flow,
The heartbeat of Earth, in its eternal glow.

But this heartbeat falters, under the weight of our deeds,
As we take more than we give, as we plant the seeds.
Of discord and harm, of imbalance and strife,
We must mend our ways, to sustain Earth's life.

So let's listen closely, to the heartbeat's plea,
To care for our planet, to set it free.
For the Earth's heartbeat is our own, a shared fate,
A call to action, before it's too late.

This poem awakens a deep connection to the vital life force of our planet, inspiring us to cherish and protect the Earth's health and balance. It reminds us of our shared responsibility to nurture the energy that sustains all life, ensuring a thriving world for future generations.

15. Earth's Voice

In the whispers of the wind, through the rustling leaves,
In the roar of the waves, on the ocean's heaves,
There's a voice that calls, with a message so clear,
It's the voice of the Earth, for those who hear.

It speaks in the songs of the birds at dawn,
In the cry of the wild, as the day is born.
It hums in the buzz of the bees at work,
In the leap of the fish, where the river quirks.

The Earth's voice is gentle, yet strong and wise,
It echoes in the canyons, and reaches the skies.
It's a voice of beauty, a voice of pain,
A voice of hope, that will always remain.

But this voice is fading, under the strain,
Of the smoke and the metal, the acid rain.
We must listen closely, and heed its plea,
To care for our planet, to set it free.

For the Earth's voice is our guiding light,
In the darkest hour, through the longest night.
It's a call to action, for me and you,
To honor the Earth, and be its stew.

This poem captures the urgent and profound beauty of Earth's natural voice, urging us to heed its call for environmental conservation. It inspires us to act with passion and commitment to protect the precious balance of our planet.

16. Echoes of the Earth

In the stillness of the world's embrace,
Echoes of the Earth trace,
A reverberation, deep and wide,
In its rhythm, we confide.

Mountains echo with ancient might,
Forests whisper through the night,
Oceans murmur with tides that turn,
In the Earth's echoes, we learn.

The rustle of leaves, the roar of falls,
Nature's call that enthralls,
Each echo, a story told,
Of the Earth's beauty, bold.

But these echoes are now faint cries,
As the balance of nature defies,
We must listen, with intent and care,
To the echoes of the Earth, everywhere.

For in these echoes, the truth is sung,
Of a planet diverse, yet one,
Let's preserve these echoes, pure and dear,
For the song of the Earth, we all revere.

This poem fosters a deep connection with the Earth, inspiring us to commit to preserving the natural echoes that resonate with the very essence of our environment. It urges us to honor and protect the harmonious rhythms that define our world, ensuring their enduring presence for future generations.

17. Embrace of the Ecosystem

In the web of life, so vast and wide,
An embrace of the ecosystem, side by side,
Interlinked lives, a network so grand,
In this embrace, together we stand.

Forests breathe with the lungs of leaves,
Oceans pulse, as the tide heaves,
Deserts whisper with sands that shift,
Mountains watch, their spirits lift.

Each creature, plant, a thread so fine,
In the ecosystem's embrace, they intertwine,
A dance of balance, a symphony of fate,
In this embrace, all life relates.

The bees that buzz from bloom to bloom,
The fungi that flourish in earth's womb,
The predator and prey in their chase,
All are held in the ecosystem's grace.

But this embrace is tender, under threat,
As human actions cause regret,
We must cherish this bond, not erase,
The beauty of the ecosystem's embrace.

So let us act with care and love,
For the earth below, the sky above,
In the embrace of the ecosystem's fold,
May we find our future, bright and bold.

This poem fosters a profound sense of unity with the natural world, inspiring a heartfelt commitment to preserving the delicate embrace of our ecosystems. It urges us to cherish and protect the intricate connections that sustain life, ensuring their enduring harmony for generations to come.

18. Gleam of the Grasslands

Across the plains where grasses sway,
A gleam of green at break of day,
The grasslands sing a silent song,
A hymn to where the wild belong.

Horizon wide, sky vast and clear,
The gleam of life is ever near,
With every blade and every flower,
The grasslands hold a timeless power.

Herds roam free, their hooves drum beats,
In this land of endless feasts,
Where bison graze and antelope leap,
The grasslands' promise, theirs to keep.

The soil rich, the air so sweet,
In this space, earth and heaven meet,
A tapestry of life, so grand,
In the gleam of the grasslands, we stand.

But this gleam is not just a sight to see,
It's the heart of nature, wild and free,
A reminder of the balance we must find,
To protect the grasslands, for all mankind.

So let us honor this emerald sea,
For in its waves, the future keys,
May the gleam of the grasslands forever shine,
A testament to design divine.

This poem ignites a deep admiration for the beautiful, life-sustaining grasslands of our world and calls us to action to preserve these vital landscapes. It urges us to protect and cherish these natural treasures, recognizing their crucial role in sustaining life and ensuring their future splendor.

19. Harmony of Habitats

In the cradle of nature, where life intertwines,
A harmony of habitats, in delicate lines.
Forests whisper secrets, deserts sing,
Oceans hum with life, in a perpetual spring.

Mountains stand tall, guardians of the sky,
Valleys cradle rivers, where waters lie.
Each habitat a verse, in nature's grand song,
Together they flourish, where they belong.

The tundra's chill, the rainforest's embrace,
Each holds a beauty, a unique grace.
From coral reefs to savannah's expanse,
Life finds a rhythm, in a timeless dance.

Yet human hands disrupt this serene,
With forests felled and waters unclean.
But hope remains in hearts that care,
To restore the balance, to heal and repair.

So let us cherish this fragile thread,
Of habitats in harmony, where life is fed.
For in their unity, we find our place,
In the grand design, of nature's grace.

This poem evokes a profound sense of awe and responsibility toward the beautiful tapestry of habitats that sustain life on our planet. It inspires us to honor and protect these vital ecosystems, recognizing their essential role in the intricate web of life that we are all a part of.

20. Harmony of the Hills

In the cradle of the hills, where whispers weave, Nature's symphony begins at dawn's reprieve.
Emerald peaks kiss the sky so blue,
In a dance of light, the morning dew.

Birdsong echoes through the ancient trees,
A melody carried by the gentle breeze.
Rivers carve their timeless path,
In harmony with the earth's gentle wrath.

Sunlight filters through the leafy green,
Painting shadows in a tranquil scene.
Flowers bloom in a riot of hues,
A testament to the life they choose.

The hills stand tall, guardians of the land,
Their silent strength, a guiding hand.
In every leaf, in every stone,
The harmony of the hills is shown.

This poem resonates with the serene and enduring spirit of the hills, capturing their timeless call to preserve the natural harmony of our world. It inspires us to heed their gentle reminder and commit to safeguarding the balance and beauty of our planet.

21. Harmony of the Horizon

Harmony of the Horizon, a line so fine,
Where the earth meets sky, in a silent sign.
A blend of colors, at dawn and dusk,
In the Horizon's harmony, a world so brusque.

The edge of day, the brink of night,
In the Horizon's balance, a serene sight.
A canvas of twilight, of first light's bloom,
In this harmonious band, no gloom.

Horizon's song, a quiet hum,
A melody of moments, to come.
The sun's slow dip, the moon's soft rise,
In the Harmony of the Horizon, time flies.

A promise of peace, in the meeting line,
Where dreams are painted, in strokes divine.
The Horizon whispers, in hues so bold,
A story of unity, of tales untold.

So let us gaze, with eyes so wide,
At the Harmony of the Horizon, our worldly guide.
For in its quiet line, a truth is known,
In the Horizon's harmony, beauty is shown.

This poem evokes a profound sense of unity and peace, inviting us to gaze upon the horizon where the boundless earth embraces the infinite sky, merging together in perfect harmony. It reminds us that, just like the horizon, we too can find harmony and oneness in the vastness of our shared world.

22. Hues of the Horizon

On the canvas of the sky, the horizon lies,
A border of hues, where the earth meets the skies,
A spectrum of colors, a visual prize,
The horizon tells tales, where the sun rises and dies.

Dawn paints with pinks, a gentle blush,
A quiet awakening, a world in hush,
The horizon whispers of the day to come,
A promise of life, where dreams are spun.

Noon glows with blues, deep and clear,
A zenith of light, for all to revere,
The horizon stands bold, a line so fine,
Dividing the earth, with a sign divine.

Dusk brings oranges, purples, and reds,
A symphony of shades, as daylight sheds,
The horizon sings of the night's embrace,
A time of rest, in the cosmic race.

But these hues are more than a sight to behold,
They're a story of the environment, untold,
A reminder of the beauty we must preserve,
For the hues of the horizon, we must conserve.

So let us cherish this daily art,
The horizon's hues, a vital part,
Of the world we share, the air we breathe,
In the hues of the horizon, we believe.

This poem awakens a sense of wonder and ignites a profound appreciation for the breathtaking artistry of our planet's horizons. It invites us to marvel at the beauty that stretches before us, reminding us of the endless possibilities and the natural splendor that surrounds us each day.

23. Lament of the Lakes and Rivers

In days of old, they sparkled bright,
Lakes and rivers, pure delight.
Their waters clear, their currents strong,
A symphony of nature's song.

But now they weep, a mournful tune,
Beneath the sun, beneath the moon.
Pollution stains their once-clear face,
A silent cry for nature's grace.

The fish that swam in joyful play,
Now struggle in the murky gray.
The reeds that whispered in the breeze,
Now stand in silence, ill at ease.

Factories spew their toxic waste,
Into the waters, with careless haste.
The lakes and rivers, once so grand,
Now bear the scars of human hand.

Yet hope remains, in hearts that care,
To heal the wounds, to mend the tear.
To cleanse the waters, pure and free,
And let the lakes and rivers be.

So let us rise, and take a stand,
To save these treasures of our land.
For in their health, we too shall find,
A future bright, for all mankind.

This poem calls for urgent action to protect and restore our lakes, highlighting their irreplaceable beauty and the vital life they sustain, urging us to preserve these natural wonders for the planet and future generations.

24. Lullaby of the Leaves

In the hush of twilight, leaves softly sway,
Whispering secrets of the fading day.
A lullaby gentle, in the evening breeze,
Sung by the rustling, whispering trees.

The forest breathes a soothing song,
A melody where we all belong.
Each leaf a note, in nature's choir,
Harmonizing with the setting fire.

The rivers murmur, the lakes reply,
Reflecting the colors of the twilight sky.
In this serene, enchanting scene,
The world finds peace, a tranquil dream.

Yet beneath this calm, a silent plea,
For us to guard this harmony.
To cherish the leaves, the air, the streams,
And protect the earth, our shared dreams.

So let us listen, and let us heed,
The lullaby of the leaves, a call to lead.
To nurture the world, with gentle hands,
And honor the beauty of these lands.

This poem evokes a serene tranquility, inviting us to embrace a deeper appreciation for the leaves and their graceful role in nature's grand symphony. It gently reminds us of the quiet beauty they bring, dancing in harmony with the wind, and the essential part they play in the delicate balance of our natural world.

25. Mountain Majesty

Amidst the heavens, where eagles dare to soar,
Stands the mountain majesty, forevermore.
A titan of the earth, with peaks so high,
Crowned with snow, against the azure sky.

The mountain's might, a fortress of stone,
A silent sentinel, regally alone.
Its slopes and valleys, a world apart,
The mountain majesty, nature's art.

The whisper of pines, the streams' soft rush,
Compose a melody, in the quiet hush.
The mountain's breath, a crisp, clean air,
A testament to beauty, beyond compare.

But this grandeur faces threats unseen,
From the march of time, and human scheme.
The mountain majesty, strong yet frail,
Calls for protection, beyond the vale.

So let us honor these giants of land,
With a heart to preserve, and a helping hand.
For the mountain majesty, in its silent plea,
Is a legacy of the earth, for you and me.

This poem captures the majestic presence of mountains, filling us with awe and reverence for their timeless beauty. It serves as a powerful reminder of our responsibility to preserve their grandeur, ensuring that these towering wonders continue to inspire and stand as symbols of strength and resilience for generations to come.

26. Murmur of the Mountains

In the stillness of dawn, where shadows play,
The mountains murmur, come what may.
Their peaks touch the sky, in silent grace,
Guardians of time, in this sacred place.

Whispers of wind through ancient pines,
Echoes of life in nature's lines.
Rivers carve their timeless path,
Through valleys deep, in nature's bath.

Snow-capped summits, pure and bright,
Reflect the sun's first morning light.
In their embrace, the world finds peace,
A harmony that will never cease.

Yet hear their murmur, a gentle plea,
To protect their beauty, wild and free.
For in their strength, they also bear,
The scars of change, the weight of care.

So let us listen, and let us heed,
The murmur of the mountains, a call to lead.
To cherish the earth, from peak to plain,
And honor the mountains, in sun and rain.

This poem stirs a profound sense of reverence for the majestic mountains, reminding us of their awe-inspiring beauty and their vital role in nurturing our planet's delicate ecosystem. It calls us to cherish and protect these towering giants, whose presence is a testament to the strength and balance of nature itself.

27. Ode to the Ozone

High above, in the stratosphere,
A guardian stands, ever clear.
The ozone layer, a shield so grand,
Protecting life across the land.

Absorbing rays from the sun's fierce light,
It keeps us safe, both day and night.
A fragile veil, yet strong and true,
In shades of blue, a wondrous hue.

But once we harmed this precious shield,
With chemicals that made it yield.
The ozone thinned, a hole appeared,
A global crisis, deeply feared.

Yet nations rose, with voices strong,
To right the past, to right the wrong.
With treaties signed and actions bold,
The ozone's tale began to unfold.

Now healing slowly, day by day,
The ozone mends in its own way.
A testament to what we can achieve,
When we unite and truly believe.

So let us honor this vital layer,
With mindful acts and constant care.
For in the ozone's gentle grace,
We find the hope for our Earth's embrace.

This poem evokes a deep sense of respect and responsibility, urging us to protect and preserve our planet's fragile ozone layer. It serves as a powerful reminder of our role in safeguarding this vital shield, inspiring us to take action to ensure a healthier, more sustainable future for all.

28. Ode to the Symphony of the Seas

Beneath the sky, where waves embrace,
A symphony of seas, in nature's grace.
The ocean's song, a timeless tune,
From dawn's first light to the rising moon.

Currents dance in a rhythmic flow,
A ballet of life, above and below.
Whales sing deep, dolphins leap high,
In the symphony of seas, beneath the sky.

Coral reefs, with colors bright,
A tapestry of life, in the ocean's light.
Fish in schools, a vibrant display,
In the underwater world, where they play.

Yet hear the plea, in the ocean's cry,
For the seas are changing, as time goes by.
Pollution spreads, and waters warm,
A call to action, to weather the storm.

So let us heed the ocean's call,
To protect and cherish, one and all.
For in the symphony of the seas, we find,
A world of wonder, beautifully intertwined.

This poem inspires a vision of harmony, where our technological achievements and the natural world coexist in perfect balance. It invites us to embrace innovation while nurturing the earth, reminding us that true progress is found in the seamless integration of human ingenuity with the beauty and wisdom of nature.

29. Pulse of the Planet

In the heartbeat of the earth, a rhythm flows,
Through forests deep and where the river goes.
Mountains echo with ancient lore,
While oceans whisper secrets from the shore.

The pulse of the planet, steady and strong,
A symphony of life, a timeless song.
From the tiniest insect to the tallest tree,
All life is connected, wild and free.

The winds carry tales of distant lands,
While deserts shift with whispering sands.
In the Arctic's chill and the tropics' heat,
The pulse of the planet never skips a beat.

Yet hear the call, a plea so clear,
To protect this world we hold so dear.
For in our hands, the power lies,
To heal the earth, beneath the skies.

So let us listen, and let us care,
For the pulse of the planet, everywhere.
In harmony with nature, let us stand,
And cherish this earth, our precious land.

This poem ignites a profound sense of unity with the natural rhythms of our planet, urging us to embrace and honor its vibrant pulse. It inspires a deep commitment to safeguarding the delicate balance of nature, encouraging us to protect and celebrate the Earth's dynamic vitality for generations to come.

30. Radiance of Rainforests

In the heart where sunlight weaves,
Emerald canopies form verdant eaves.
Whispers of life in every leaf,
A symphony of green, beyond belief.

Mighty trees reach for the sky,
Their crowns where birds and breezes fly.
Beneath, a world both damp and warm,
Where nature's wonders take their form.

Rainfall dances on the ground,
A rhythmic pulse, a soothing sound.
Nourishing roots that delve so deep,
In the rainforest's embrace, secrets keep.

Colors burst in vibrant hues,
Flowers bloom with morning dews.
Creatures great and small reside,
In this lush, life-giving tide.

The air is thick with fragrant breath,
A testament to life's depth.
From the forest floor to the canopy high,
A living tapestry meets the eye.

Yet fragile is this radiant land,
At the mercy of a human hand.
Protect these realms of green and gold,
For their stories must forever be told.

This poem captures the vibrant essence of rainforests, celebrating their vital role in sustaining the health of our environment. It illuminates the rich tapestry of life they support and inspires us to protect these lush, life-giving ecosystems, recognizing their indispensable contribution to our planet's well-being.

31. Reflections in the Rain

In the gentle weep of heaven's domain,
The earth is washed, anew again.
Each droplet holds a mirror, a frame,
Capturing life's moments, in reflections of the rain.

The streets turn to rivers, of silver and gray,
Reflecting the world in a watery array.
A canvas in motion, a fluid terrain,
Life's fleeting images, in reflections of the rain.

People pass by, umbrellas in hand,
Their stories ripple, across the wet land.
A momentary glimpse, then gone just the same,
In the transient beauty, of reflections in the rain.

So let us ponder, in this liquid grain,
The reflections we see, what do they explain?
For in each puddle's depth, a truth may pertain,
A piece of ourselves, in reflections of the rain.

This poem evokes the contemplative mood that a rainy day can bring, with its reflective surfaces and the introspective atmosphere it creates.

32. Reflections in the River

In the river's gentle flow,
Mirrored worlds begin to show.
Sky and trees in liquid dance,
A tranquil, fleeting, soft romance.

Ripples whisper tales untold,
Of life beneath the surface cold.
Fish dart through the dappled light,
In this realm of day and night.

Sunlight kisses water's face,
Creating patterns, pure and grace.
Reflections shift with every breeze,
A living canvas, nature's ease.

Yet, beneath this beauty bright,
Lies a call for us to fight.
To keep these waters pure and clear,
For future generations dear.

Pollution's shadow, creeping slow,
Threatens all that rivers know.
We must act, protect, and strive,
To keep these vital streams alive.

So let us cherish every gleam,
In rivers' reflections, let us dream.
Of a world where nature thrives,
And every river's beauty survives.

This poem ignites a deep sense of stewardship for our rivers, celebrating their crucial role as reflections of our environment's health. It encourages us to honor and protect these flowing lifelines, recognizing their essential contribution to the vitality and balance of our natural world.

33. Reflections of the Rain

Reflections of the Rain, a mirror of the sky,
A canvas of droplets, from clouds so high.
Each one a prism, a dance of light,
In the rain's soft reflections, the world's alight.

The pitter-patter, a rhythmic beat,
On leaves, on earth, a symphony sweet.
The rain whispers tales of life anew,
In its gentle reflections, the world's askew.

It paints the air with a fresh, clean scent,
A natural cleanse, a heavenly sent.
The Reflections of the Rain, so clear, so bright,
In each droplet, a bit of night.

Yet, the rain speaks of cycles, of ebb, of flow,
Of the balance of nature, of growth, of woe.
For in its reflections, a truth is found,
In the rhythm of rain, the earth is bound.

So let us gaze in the rain's soft mirror,
For in its depths, life's clearer.
The Reflections of the Rain, a lesson to heed,
In its simple beauty, our planet's creed.

This poem reflects the cleansing and life-giving essence of rain, and its role in the ever-turning wheel of the environment.

34. Rivers of Hope

In the heart of the land, where the waters flow,
Run the rivers of hope, in a gentle glow.
A lifeline, a promise, in each silver stream,
Carrying dreams of tomorrow, in its ceaseless gleam.

The rivers wind and bend, through valley and dale,
Telling tales of the earth, an ageless tale.
A dance of droplets, under the sun's warm cope,
The rivers of hope, a symbol of our hope.

They nurture the fields, quench the thirsty land,
A touch of life, from nature's hand.
The rivers of hope, in their quiet might,
Are a beacon of change, in the fading light.

But these rivers face threats, from the choices we make,
Pollution and drought, a high stake.
We must guard these waters, keep them pure and clear,
For the rivers of hope, are treasures dear.

So let's pledge to the rivers, in their graceful sweep,
To protect their journey, to the oceans deep.
For the rivers of hope, in their song so bright,
Carry the future, in their flight.

This poem ignites a profound sense of stewardship for our rivers, highlighting their essential role in sustaining life and nurturing hope for a sustainable future. It calls us to cherish and protect these vital waterways, recognizing their power to sustain both our environment and our aspirations for a thriving, harmonious world.

35. Seaside Serenade

Upon the shore where waves caress the sand,
A melody plays, a Seaside Serenade so grand.
The ocean's rhythm, a symphony so true,
Sings of mysteries deep, in shades of blue.

The seagulls join in chorus, with cries that soar,
Above the crashing waves, above the ocean's roar.
The shells whisper secrets, of the deep sea's lore,
In the Seaside Serenade, a song of yore.

The salt in the air, the breeze on one's face,
Compose a tune of freedom, of wild, untamed grace.
The horizon stretches, an endless escapade,
In the embrace of the sea, the Seaside Serenade.

As the sun sets low, in a fiery display,
The sky joins the concert, in its own unique way.
With colors that blend, in a heavenly parade,
The day ends in beauty, with the Seaside Serenade.

So let the waves carry you, in their rhythmic dance,
In the Seaside Serenade, find romance.
For in each ebb and flow, memories are made,
In the timeless song of the sea, the Seaside Serenade.

This poem captures the enchanting allure of the seaside, celebrating the rhythmic beauty of its natural serenade. It immerses us in the soothing melody of the ocean's waves and the gentle caress of the sea breeze, inviting us to revel in the tranquil majesty and timeless charm of this coastal paradise.

36. Seasons of Change

Spring awakens with a gentle sigh,
Blossoms bloom, and birds take to the sky.
Rivers swell with melting snow,
Life returns in a vibrant show.

Summer blazes with a golden hue,
Fields of green and skies of blue.
Yet heatwaves scorch the thirsty land,
A reminder of nature's fragile hand.

Autumn whispers with a cooling breeze,
Leaves turn gold and fall from trees.
Harvests gathered, days grow short,
A time for reflection and nature's report.

Winter blankets with a frosty veil,
Snowflakes dance in a silent tale.
But warmer winters now we see,
A sign of change, a call to be.

Each season tells a story clear,
Of beauty, balance, and what we hold dear.
Yet climate shifts and patterns break,
A world at risk, for our own sake.

Let us heed the seasons' call,
To cherish, protect, and care for all.
For in their cycles, life sustains,
A precious gift, in sun and rains.

This poem inspires a deep respect for the natural cycles of our planet, emphasizing the urgent need to act in protecting our environment amidst a changing climate. It calls us to honor the delicate balance of nature and motivates us to take meaningful steps to safeguard the Earth's precious ecosystems for future generations.

37. Serenade of the Shoreline

Along the edge where land meets sea,
A serenade plays, a melody free,
The shoreline sings in waves and sand,
A symphony crafted by nature's hand.

The tide rolls in with a soft caress,
In its touch, a power to bless,
It sculpts the shore with every wave,
A rhythm of giving, a rhythm to save.

Seashells whisper tales of the deep,
Of moonlit waters where mysteries sleep,
Each one a note in the ocean's choir,
A piece of the puzzle, a spark of fire.

The breeze carries the salt of the sea,
A breath of life, wild and free,
It dances with dunes, a delicate waltz,
In the serenade of the shoreline, time halts.

But this border of beauty, so fragile and fine,
Faces threats from the hands of time,
We must listen to the shoreline's plea,
To protect its song for eternity.

So let us join in this coastal hymn,
To honor the shore, at the water's brim,
For in the serenade of the shoreline's lore,
Lies a message of hope, worth fighting for.

This poem sparks a profound sense of wonder and a powerful call to action to protect the beautiful, life-sustaining shorelines of our world. It invites us to marvel at their splendor and motivates us to take decisive steps to preserve these vital, vibrant edges of our planet for the well-being of all life.

38. Shadows of the Shore

Where the waves kiss the land, and the tides rise and fall,
The shadows of the shore whisper a timeless call.
A dance of light and dark, on the sand's soft bed,
Telling tales of the deep, in the hues of blue and red.

The sun sets, painting the horizon in gold,
As the shore's silhouette, in the evening, unfolds.
The sea's gentle lapping, a rhythm so pure,
Plays with the shadows, in allure.

In the quiet of dusk, the shore's shadows grow,
Stretching long fingers, as the cool breezes blow.
They reach for the ocean, in a tender embrace,
A moment of beauty, nature's grace.

But these shadows tell, of a story untold,
Of the life beneath waves, both fearless and bold.
Of coral and creatures, that we must protect,
For the shadows of the shore, they reflect.

Let us honor the whispers, of the shadows we see,
By preserving the oceans, keeping them free.
For the shadows of the shore, in their silent plea,
Remind us to care, for the vast, living sea.

This poem fosters a profound appreciation for the delicate harmony between land and sea, highlighting the crucial importance of environmental conservation. It encourages us to cherish this vital balance and inspires us to commit to protecting our planet's natural beauty and ecological integrity.

39. Silent Echoes

In the hallowed halls of the heart's keep,
Where secrets lie and memories seep,
There resonate, without a peep,
The silent echoes, in a leap.

These echoes, they are not of sound,
But of moments lost, then found.
They reverberate, without a bound,
In the soul's depths, where they're crowned.

A laugh, a tear, a touch, a glance,
Each leaves behind its own resonance.
A silent echo, a subtle dance,
Of life's intricate, vast expanse.

So listen to the quiet, the still,
For silent echoes, with meaning, fill.
They are the whispers of the will,
The echoes that the silence spill.

This poem beautifully captures the profound and often unspoken echoes of our experiences and emotions, revealing the deep, resonant impact they have on our inner selves. It invites us to explore and embrace the subtle yet powerful ripples of our journey, connecting us to the deeper truths of our shared human experience.

40. Song of the Savanna

The savanna sings a timeless tune,
Under the watchful eye of the moon,
A symphony of life in golden grass,
Where moments whisper and memories pass.

Acacias rise like notes on a line,
Silhouetted against the sky so fine,
Elephants trumpet, lions roar,
Each sound a verse in the savanna's lore.

Giraffes browse the treetops high,
Their patterns a part of the earth and sky,
Zebras graze in the sun's warm glow,
A dance of stripes, a living show.

The savanna breathes with the wind's caress,
A canvas of survival and success,
Its song is one of balance and grace,
In this open, wild, and sacred place.

But the savanna's tune is under threat,
As human pressures and dangers beget,
We must listen to its ancient call,
And protect its harmony, once and for all.

For the song of the savanna is our own,
A reminder of the seeds we have sown,
Let's join in chorus with this land so vast,
And make the song of the savanna last.

This poem vibrates with the spirit of the savanna, stirring a heartfelt commitment to preserving its unique and vibrant ecosystem. It invites us to connect deeply with the rich tapestry of life that defines this extraordinary landscape and inspires us to protect and cherish its irreplaceable beauty for future generations.

41. Song of the Snowcaps

Song of the Snowcaps, a melody pure,
Echoing silence, in white allure.
On peaks so high, where eagles soar,
The snowcaps sing, a mythic lore.

In the realm of ice, where time stands still,
The Song of the Snowcaps, a winter's quill.
Writing tales of frost, of nights so crisp,
A world of wonder, in an icy wisp.

The mountains' crowns, so old, so wise,
Beneath the glow of aurora skies.
They whisper of balance, of a world interlaced,
In the snowcaps' song, the earth's embraced.

Yet, they weep for warmth that comes too soon,
For the snow that fades 'neath the sun of noon.
The Song of the Snowcaps, a warning clear,
For the changing climes, the snowcaps fear.

Let us hear their tune, so soft, so bold,
In the silence of snow, the future's told.
To guard the cold, the ice, the frost,
For without the snowcaps, much will be lost.

So let the Song of the Snowcaps ring,
In their crystal notes, our actions sing.
To preserve the chill, the white, the grace,
For the snowcaps' song, in our embrace.

This poem resonates with the pristine beauty and crucial significance of our planet's snowcapped peaks, serving as a poignant reminder of the delicate balance we must uphold. It inspires us to revere these majestic heights and renew our commitment to preserving their fragile splendor for the well-being of our planet.

42. Symphony of Nature

Nature's orchestra begins to play,
At dawn's first light, a new day.
Birds sing melodies so sweet,
In harmony, their voices meet.

The wind whispers through the trees,
A gentle rustle, a calming breeze.
Leaves dance to the rhythm of the air,
A ballet of green, beyond compare.

Rivers flow with a soothing sound,
A symphony of life all around.
Fish leap in a sparkling arc,
In waters clear, from dawn till dark.

Mountains stand as silent guards,
Echoing the songs of bards.
Their peaks touch the sky so high,
A testament to nature's sigh.

Yet, within this symphony so grand,
Lies a plea from the land.
To protect, preserve, and cherish,
So this music never perishes.

Climate shifts and seasons change,
A reminder of nature's range.
Let us tune our hearts and minds,
To the symphony that nature binds.

This poem ignites a deep love for the natural world and fuels a passionate desire to protect its breathtaking, yet delicate, symphony. It calls us to embrace the beauty of nature's harmonious rhythms and inspires us to safeguard the intricate melodies that sustain our planet's wonder.

43. Symphony of Rain

In the hush of twilight's veil,
Raindrops start their gentle tale.
A symphony begins to play,
In nature's most enchanting way.

Each droplet sings a note so clear,
A melody for all to hear.
Pitter-patter on the leaves,
A soothing sound that never grieves.

Rivers swell with every beat,
A rhythmic pulse beneath our feet.
Streams and brooks join in the song,
A flowing chorus, pure and strong.

The earth drinks deep this liquid gold,
A gift from skies, both young and old.
Plants awaken, roots embrace,
The life-giving rain's embrace.

Yet, within this symphony's grace,
Lies a plea for our embrace.
To cherish, guard, and understand,
The delicate balance of this land.

For rain sustains the world we know,
From mountain peaks to valleys low.
Let us honor every drop,
And ensure this symphony never stops.

This poem captures the soul-stirring melody of the rain and its profound, transformative impact on the world around us. It invites us to listen to the rain's gentle symphony, appreciating its power to renew and invigorate the natural world, and inspiring us to cherish and protect this vital force of nature.

44. Symphony of Soil

Beneath our feet, a world unseen,
Where life begins in shades of green.
A symphony of roots and earth,
The soil, a cradle of life's birth.

Microbes dance in hidden light,
Turning decay into new life.
Nutrients flow in silent streams,
Feeding plants and shaping dreams.

Worms weave through the darkened loam,
Creating pathways, making homes.
Each grain of soil, a story told,
Of ancient times and secrets old.

Rainfall seeps into the ground,
A gentle rhythm, a soothing sound.
Binding soil in nature's embrace,
A cycle of life, a sacred space.

Yet, fragile is this earthen song,
At risk from harm, from doing wrong.
Pollution, erosion, human greed,
Threaten the soil on which we feed.

Let us honor this precious ground,
Protect its health, keep it sound.
For in the soil, life's symphony,
Plays on in perfect harmony.

This poem highlights the vital role of soil conservation and celebrates the intricate beauty of this often-overlooked cornerstone of our ecosystem.

45. Symphony of the Seas

In the vast expanse where blue meets blue,
The ocean sings a song so true.
Waves crash and whisper on the shore,
A symphony of nature, forevermore.

The tides, they dance to the moon's soft call,
Rising high and gently fall.
Currents weave a tale untold,
In waters deep and stories old.

Coral reefs, a vibrant choir,
Colors blaze like living fire.
Fish dart in a ballet grand,
A testament to nature's hand.

Yet, beneath this beauty lies,
A plea for help, a world that cries.
Pollution stains the waters clear,
A silent threat, a growing fear.

Let us heed the ocean's plea,
Protect its depths, keep it free.
For in its waves, life's symphony,
Plays on in perfect harmony.

This poem inspires us to forge a harmonious bond between our technological advancements and the natural world, embracing the beauty of their synergy.

46. Symphony of the Seasons

Spring awakens with a gentle song,
Blossoms bloom, and days grow long.
Birds return with melodies sweet,
In nature's symphony, life's heartbeat.

Summer blazes with a vibrant tune,
Sunlight dances, warm afternoons.
Fields of green and skies so bright,
A season of joy, pure delight.

Autumn whispers with a mellow tone,
Leaves turn gold, a world overgrown.
Harvests gathered, colors ablaze,
Nature's canvas, a fiery haze.

Winter arrives with a hushed refrain,
Snowflakes fall, a silent gain.
Blanketing earth in tranquil white,
A serene pause, a peaceful night.

Each season sings its unique part,
A symphony that stirs the heart.
Yet climate shifts and changes loom,
A call to act, before the gloom.

Let us cherish every note,
Protect the earth, keep it afloat.
For in this symphony of time,
Nature's beauty is truly sublime.

This poem celebrates the cyclical splendor of the seasons and the harmonious rhythm they weave into the fabric of our world, reminding us of nature's timeless dance.

47. Tears of the Earth

Beneath the sky's vast, watchful eye,
The Earth lets out a somber sigh.
Her tears, they fall, not seen but felt,
In every blow that nature's dealt.

From forests deep to oceans wide,
Her silent sobs, she cannot hide.
The tears of Earth, for all her children lost,
For every line we've carelessly crossed.

She weeps for trees, once tall and grand,
Now fallen leaves upon the sand.
For skies once clear, now veiled in grey,
Her tears, they speak, in their silent way.

The Earth, she mourns in rivers clear,
That run with sorrow and with fear.
For ice caps melting, seas that rise,
Her teardrops glisten in our eyes.

Yet, in her weeping, there's a plea,
A call to you, a call to me.
To dry her tears, to heal her pain,
To bring her back to life again.

So let us listen, let us start,
To mend the tears, to heal her heart.
For in her cries, the Earth does yearn,
For love returned, for tides to turn.

This poem illuminates the Earth's delicate fragility and our shared duty to nurture and safeguard our planet, urging us to unite in its protection and care.

48. The Blooming Biome

In the heart of nature's grand design,
Lies a biome, lush and fine.
A tapestry of life, so vast,
From ancient roots to skies so vast.

Flowers bloom in vibrant hues,
A symphony of reds and blues.
Insects buzz and birds take flight,
In this realm of pure delight.

Trees stand tall, their canopies wide,
Sheltering life on every side.
Leaves whisper secrets to the breeze,
A dance of green among the trees.

Rainfall nourishes the ground,
A rhythmic pulse, a soothing sound.
Streams and rivers carve their way,
Through this blooming, bright display.

Yet, fragile is this wondrous land,
At the mercy of a human hand.
Pollution, deforestation's blight,
Threaten this biome's radiant light.

Let us cherish, guard, and care,
For this blooming biome, beyond compare.
For in its beauty, life sustains,
A precious gift, in sun and rains.

This poem celebrates the vibrant, interconnected beauty of the Blooming Biome, a vivid reminder of the precious diversity of life that we are entrusted to protect and cherish.

49. The Blossoming Bog

The Blossoming Bog, a vibrant quilt,
Where flowers bloom and life is built.
A patchwork of petals, in hues so bright,
In the bog's soft glow, a pure delight.

Amidst the marsh, where waters seep,
The Blossoming Bog does secrets keep.
A haven for frogs, for birds, for bees,
In its lush embrace, a gentle ease.

The air is thick with the scent of bloom,
Where nature's palette does resume.
In every nook, in every cranny,
The bog's life thrives, so uncanny.

Yet, the bog sings a song of care,
A reminder that all is rare.
For in its depths, a world unseen,
In the Blossoming Bog, life's serene.

So let us treasure this floral bed,
Where the earth's soft whispers are said.
For in the bog, so wild, so free,
Lies a beauty, for you and me.

This poem captures the essence of the bog's vibrant ecosystem, celebrating its unique beauty and the intricate web of life it sustains, urging us to honor and protect this remarkable sanctuary.

50. The Bountiful Basin

In the lap of the earth, where waters embrace,
Lies the Bountiful Basin, a nurturing place.
A cradle of life, where the streams converge,
A symphony of abundance, nature's urge.

Verdant banks kiss the basin's brim,
As life within stirs at every whim.
Fish leap with joy, birds in chorus rise,
A celebration of earth, under open skies.

The basin, a bowl of plenty untold,
With stories of sustenance, from days of old.
A giver of life, a receiver of rain,
In the cycle of life, it shall remain.

But now the basin, so rich and wide,
Feels the pangs of a changing tide.
The hands of time and toil weigh,
On the bountiful gifts, day by day.

Let us cherish the basin, its gift so rare,
By protecting its essence, with utmost care.
For in its depths, a world does thrive,
In The Bountiful Basin, we keep hope alive.

This poem kindles a deep appreciation for vital ecosystems like basins, which nurture an incredible diversity of life. It calls us to recognize their significance and commit to safeguarding them for the future.

51. The Celestial Canopy

The Celestial Canopy, a vault so grand,
A tapestry woven by nature's hand.
A dome of blue, of cloud, of light,
Where day meets night in an endless flight.

Above the earth, it stretches wide,
A sheltering cloak, the world's outside.
In its expanse, the sun does roam,
The moon and stars find their home.

The Canopy sings of the air we breathe,
Of the life it holds, the web it weaves.
A guardian of all, so high, so vast,
In its embrace, the earth is cast.

Yet, it cries for the care it needs,
For the wounds it bears, where neglect leads.
The Celestial Canopy, in its silent plea,
Asks us to look up, to see, to free.

Let's honor the sky, the atmosphere,
For in its health, our path is clear.
To cherish the Canopy, to make amends,
For on its grace, life depends.

So let us sing with the Canopy's tune,
Under the sun, the stars, the moon.
To protect the blue, the celestial sweep,
For the Canopy's song, is ours to keep.

This poem ignites a sense of awe and responsibility toward the vast, beautiful sky that envelops our planet, reminding us of the critical importance of preserving our atmosphere for future generations.

52. The Drowsy Deltas

In the arms of the earth, where rivers end,
The Drowsy Deltas softly blend.
A cradle of life, in silty beds,
Where waters slow, and greenery spreads.

Gentle giants, the deltas lie,
Under the watch of the open sky.
Their fingers stretch to the sea's embrace,
In a sleepy yawn, at a languid pace.

Here, the reeds whisper secrets old,
Of silt and sediment, sun and cold.
A mosaic of channels, both deep and shallow,
Harboring life, in each quiet billow.

The deltas dream in shades of blue and green,
A haven for creatures, often unseen.
Fish dart below, birds soar above,
In this nursery of nature, woven with love.

But hush, the deltas now face a plight,
As tides of change blur their sight.
The waters murmur with a troubled frown,
For the hand of man weighs heavy down.

Let us sing a lullaby for these sleepy heads,
Preserve their peace, their watery beds.
For the Drowsy Deltas, in their gentle rest,
Hold the key to life, at its very best.

This poem beautifully captures the serene yet vulnerable essence of the deltas, highlighting the urgent need for their conservation and our responsibility to protect these precious landscapes.

53. The Dying Desert

A sea of sand, once teemed with life,
Now faces a future of strife,
The dying desert, whispers its plea,
A cry for help, an urgent decree.

Dunes that danced in the wind's embrace,
Now stand still, a barren place,
Where once the cactus flower bloomed,
Silence and emptiness have loomed.

The sun beats down with a scorching frown,
On cracked earth, browned and down,
The desert's breath, a heated sigh,
Under the vast, unyielding sky.

Oases shrink, wells run dry,
As creatures search for relief, and sigh,
The balance of life has gone awry,
In the dying desert, under the eye.

But this desert's tale is not yet done,
For in its sands, there's a battle to be won,
To revive the life that once did thrive,
To ensure the desert's tale survives.

So let us heed the desert's call,
To restore its glory, for one and all,
For in the grains of its ancient sand,
Lies a story of survival, stark and grand.

This poem shines a light on the plight of desert environments, calling us to action with a sense of urgency and dedication to conserve and protect these extraordinary and fragile landscapes.

54. The Earth Song

Hear the Earth Song, a melody profound,
A chorus of life, where hope is found.
It's the rustling leaves, the roaring falls,
Nature's anthem, to which the heart calls.

The Earth Song plays, from mountain to sea,
A tune of unity, for you and me.
It's the whisper of grains, in fields of gold,
A story of harmony, ages old.

The rhythm of wings, in the azure sky,
The Earth Song soars, where eagles fly.
It's the blooming buds, the buzzing bees,
A symphony of peace, in the gentle breeze.

But this song is fading, under strain,
Muted by the noise, of greed and pain.
We must rise, and lend our voice,
To sing the Earth Song, make the choice.

For the Earth Song is a call to care,
To cherish the world, clean and fair.
It's a plea for action, a demand for love,
A hymn for the planet, we're a part of.

So let's join the chorus, of the Earth Song,
And heal the world, where we belong.
For in its melody, we find our role,
To save the Earth, to make it whole.

This poem fosters a profound connection with the natural world, inspiring a steadfast commitment to environmental stewardship and a passion for preserving our planet's wonders.

55. The Earth Weeps

Upon the canvas of the world, a tear falls, silent, deep,
From the eyes of the skies, the Earth begins to weep.
A lament for the lost, for the beauty we forsake,
For the scars we leave behind, with every step we take.

The Earth weeps in the rivers, polluted and forlorn,
In the smog-filled skies, where once the day was born.
It weeps in the forests, where stumps stand in lieu,
Of ancient trees that in the gentle breezes blew.

The weeping is a whisper, a sorrowful refrain,
A plea for mercy, to ease the planet's pain.
For every drop that falls, from the Earth's weary eyes,
Is a call to the living, to heed the silent cries.

Yet, within the weeping, there's a spark of hope that lies,
In the actions of the caring, under the same vast skies.
For the tears of the Earth can be wiped away,
With hands that heal, with those who'll sway.

So let us be the comfort, in the Earth's time of need,
Let's nurture and cherish, let's plant the seed.
For the Earth weeps not for itself, but for us, its kin,
In our hands lies the power, a new start to begin.

This poem underscores the urgency of environmental challenges while kindling hope through the power of collective action, inspiring us to unite in healing and rejuvenating our planet.

56. The Echoing Forest

In the heart of the echoing forest,
Where whispers of leaves are a chorus,
The trees stand tall, guardians of lore,
Roots deep as secrets, reaching the core.

A canopy woven with emerald thread,
Sunlight and shadow in a dance overhead,
The forest breathes, a symphony serene,
Inhaling the world, exhaling the green.

Each tree, a pillar of life's grand tent,
A testament to years silently spent,
In the service of all, giving more than they take,
For the forest is alive, make no mistake.

The rustling leaves tell tales of old,
Of cycles of life, both gentle and bold,
They speak of balance, of give and of get,
A harmony we must not forget.

For the echoing forest is more than wood,
It's the earth's very lungs, its mood,
A refuge for creatures great and small,
A reminder that we are part of it all.

So let's cherish these woods, these silent sentinels,
For in their survival, lies ours as well,
May the echoing forest continue to thrive,
For as long as it stands, we are alive.

This poem resonates with the spirit of the forests, celebrating their vital role in sustaining our planet and inspiring us to cherish and protect these essential green sanctuaries.

57. The Edge of Dawn

At the edge of dawn, where night meets the day,
A sliver of light cuts through the gray.
The horizon blushes, the stars fade away,
As the world awakens to the sun's first ray.

The sky, a canvas of soft pastel hues,
Whispers of morning, in pinks and blues.
The edge of dawn, a time of renew,
When dreams of the night find their morning's cue.

Birds sing their chorus, a melody born,
In the gentle hours of the newborn morn.
The air is fresh, the earth adorned,
With dew-kissed flowers at the edge of dawn.

So let us rise with this time so rare,
And breathe in the promise hanging in the air.
For at the edge of dawn, hope is drawn,
In the quiet beauty of the breaking dawn.

This poem captures the serene and uplifting essence of dawn's first light, infusing us with hope and the promise of new beginnings.

58. The Eternal Ecosystem

The Eternal Ecosystem, a dance of life,
In its endless rhythm, there's no strife.
A circle of birth, of growth, of rest,
In nature's arms, we're truly blessed.

From the tiniest seed to the mightiest tree,
Each has a role, a part to be.
In the Eternal Ecosystem's grand design,
Every creature, every plant, a line.

The buzz of bees, the leap of deer,
In this system, they all draw near.
A balance of give, of take, of share,
In this cycle, all is fair.

The waters flow, the mountains rise,
Under the vast, embracing skies.
In the Ecosystem, life's tapestry weaves,
A story of earth, of air, of leaves.

Yet, it whispers of care, of a tender need,
For the hand of man, not to harm, but to heed.
The Eternal Ecosystem, fragile and strong,
In its harmony, we all belong.

So let us sing, with voices wide,
For the Ecosystem, our global pride.
To preserve, to protect, to admire,
The Eternal Ecosystem, our world's fire.

This poem inspires a deep sense of unity and responsibility toward the intricate and enduring web of life that is our planet's ecosystem, urging us to embrace our role in nurturing and preserving this precious balance.

59. The Fragile Forest

The Fragile Forest whispers low,
A verdant realm where soft winds blow.
A canopy of emerald lace,
A delicate, enchanting place.

Each leaf, a note in nature's hymn,
In the forest's light, so dim.
The trees stand tall, yet frail and slight,
Guardians of day, sentinels of night.

The Fragile Forest, a living mesh,
Where life weaves through each crevice fresh.
A habitat so diverse, so vast,
In its quiet strength, a spell is cast.

But oh, how delicate the balance here,
As threats encroach, both far and near.
The axe, the flame, the clearing's toll,
Upon this forest, they take their roll.

The Fragile Forest, in silent plea,
Calls to us, to you and me.
To protect, to cherish, to revere,
This haven of life we hold so dear.

For if we lose this fragile trove,
We lose a treasure from above.
So let us stand, with hearts so bold,
To save the forest, more precious than gold.

This poem serves as a gentle reminder of the vulnerability and intrinsic value of our forests, urging us to act with care and respect towards these natural treasures.

60. The Glowing Grove

In the hush of twilight, where fireflies roam,
Lies a wonder unseen, the Glowing Grove's home.
A symphony of light, in the forest's heart,
Where bioluminescent trees do their part.

Each leaf a lantern, each branch a beam,
Casting a glow, like a living dream.
The grove, it dances in radiant attire,
A natural marvel that sparks desire.

The night comes alive with each gentle sway,
A ballet of brilliance, night's own bouquet.
Creatures of dusk, in shadows they thrive,
Amongst glowing boughs, they come alive.

But this grove, so rare, faces twilight's end,
As human touch, its fibers bend.
The glow dims under the strain of blight,
A plea for care, in the fading light.

Let's preserve the glow, the magic we've found,
In the heart of the woods, where light abounds.
For the Glowing Grove, a treasure to behold,
Speaks of mysteries, centuries old.

This poem captures the enchanting beauty of a glowing grove, illuminating the need to safeguard these magical sanctuaries and preserve their wonder for future generations.

61. The Graceful Grassland

The Graceful Grassland, waves in the breeze,
A sea of green, with ease it frees.
Rolling meadows, under the sky's expanse,
Where the blades of grass do sway and dance.

A carpet of life, so soft, so vast,
In the Graceful Grassland, time is surpassed.
Each tuft and tassel, a stroke of art,
In this tranquil land, pure of heart.

The hum of life, a subtle sound,
In the Grassland's grace, peace is found.
Herds roam free, in gentle troves,
Through the golden plains and verdant groves.

The Graceful Grassland, a haven wide,
Where the earth's joys do not hide.
A symphony of hues, from gold to green,
In this serene scape, beauty's seen.

Beneath the sun, the grassland sings,
Of cycles, seasons, nature's rings.
A dance of life, in harmony,
The Grassland's grace, for all to see.

So let us tread with care, with love,
For the Graceful Grassland, under stars above.
A treasure to cherish, to hold dear,
For in its grace, the world's cheer.

This poem embodies the elegance and serenity of the grasslands, reminding us of their gentle yet profound impact on the health and beauty of our planet, and inspiring us to cherish and protect these vital landscapes.

62. The Green Canopy

Beneath the vast expanse of the sky's embrace,
Lies the green canopy, nature's grace.
A shelter of leaves, a tapestry so wide,
Where the whispers of the earth reside.

The green canopy breathes, a life of its own,
A crown of the forest, where seeds are sown.
A haven for birds, a home for the light,
Filtering through the leaves, a sight so bright.

The rustle of foliage, in the gentle breeze,
Tells a story of balance, of peace and ease.
The green canopy stands, in quiet defiance,
A testament to life's resilient alliance.

But this verdant roof, this ceiling so high,
Faces threats that make the spirits sigh.
We must protect this canopy, so lush, so alive,
For under its arch, the world's wonders thrive.

So let's raise our voices, in a chorus so green,
To preserve the canopy, to keep it serene.
For the green canopy, in its leafy sprawl,
Is a reminder of beauty, that belongs to us all.

This poem inspires a profound sense of wonder and responsibility toward the lush canopies that adorn our planet, urging us to honor and protect these vital green treasures.

63. The Hidden Hills

In the shroud of mist, where secrets dwell,
Stand the Hidden Hills, in silence they tell.
A veil of green, draped o'er their face,
Guarding the wonders of a forgotten place.

These hills, they whisper through the trees,
Of ancient paths and rustling leaves.
A refuge for the wild, a haven so still,
In the embrace of the Hidden Hills.

The brooks that babble, the birds that call,
In this secluded realm, they enthrall.
A tapestry of life, woven unseen,
In the folds of the hills, so serene.

But beneath the calm, a shadow looms,
As progress marches, and nature's room
Grows ever smaller, the hills recede,
A silent plea, their only creed.

Let us seek the hills, so hidden and true,
And protect their legacy, for me and you.
For in their quiet, the earth's heart beats,
In the Hidden Hills, where mystery greets.

This poem unveils the serene beauty and vital importance of preserving the hidden natural landscapes around us, inspiring us to safeguard these quiet treasures that sustain our world.

64. The Infinite Isles

Beyond the horizon, where dreams take flight,
Lie the Infinite Isles, bathed in light.
A tapestry of green, blue, and gold,
Stories of the earth, silently told.

Isles without number, scattered like jewels,
Cradled by waves, nature's precious tools.
Each a world apart, yet together they sing,
A chorus of life, in the ocean's ring.

Palm fronds sway to the sea's gentle kiss,
Sands whisper tales of ecological bliss.
Coral reefs bloom in vibrant array,
A kaleidoscope of colors at play.

But these isles face a future unsure,
As seas rise and climates endure.
The infinite becomes finite in our hands,
As we shape the fate of these fragile lands.

Let us be the keepers of isle and sea,
To preserve their infinity for posterity.
For in their boundless beauty we find,
A mirror of the best of humankind.

This poem ignites a profound sense of stewardship for the beautiful yet vulnerable island ecosystems around the world, urging us to cherish and protect these precious havens with unwavering dedication.

65. The Last Leaf

Upon a branch where autumn's breath did cleave,
There clung in quiet defiance, the last, lonely leaf.
A remnant of summer's vibrant sheaf,
Holding fast to life, in belief.

It watched its kin, in spirals leave,
To join the earth, a tapestry to weave.
Yet there it stayed, a single motif,
A symbol of persistence, the last leaf.

The wind did howl, the nights grew thief,
As winter's chill brought its cold grief.
But still it hung, an emblem brief,
Of time's relentless march, the last leaf.

Until one morn, to the ground's relief,
It fluttered down, in a dance so chief.
And there it lay, a chapter's end, not in mischief,
A testament to change, the last, honored leaf.

This poem captures the poignant beauty and resilient spirit of the last leaf of autumn, celebrating its enduring grace and inspiring us to find strength and hope in life's fleeting moments.

66. The Last River

In the heart of a dying land, it flows,
A silver thread where life once rose.
Mountain's weep and forests sigh,
As the Last River whispers goodbye.

Once a torrent, fierce and wild,
Now a stream, gentle and mild.
Its waters clear, yet filled with tears,
Echoing tales of forgotten years.

The fish that danced in joyous play,
Have long since vanished, gone away.
Birds that sang from dawn till night,
Now silent, lost in endless flight.

The trees that lined its verdant banks,
Stand bare and broken, giving thanks.
For every drop that still remains,
In this river's ancient veins.

The sun sets low, a crimson hue,
Reflecting dreams that once were true.
And as the stars begin to gleam,
The Last River flows through a final dream.

A dream of life, of hope, of green,
Of days when nature's hand was seen.
But now it fades, a ghostly shiver,
The end is near for the Last River.

This poem stirs a deep sense of urgency and responsibility, urging us to protect and preserve the last precious treasures of our natural world with unwavering dedication and care.

67. The Last Standing Tree

In a land stripped bare, under a sky so wide,
Stands the last tree, with nowhere to hide.
A sentinel of time, a keeper of tales,
Its branches still reach, though the greenery pales.

The last standing tree, a monument of might,
Against the winds of change, it stands upright.
A symbol of hope, in a world turned gray,
A reminder of nature, that we've cast away.

Its leaves may be few, its bark scarred and old,
But the story it tells is brave and bold.
Of forests that once clothed the earth in green,
Of rivers that sparkled, so clear and pristine.

The last standing tree, alone in its fight,
Echoes the whispers of the forest's plight.
A call to action, from the roots to the crown,
To turn the tide, to not let it drown.

So let us gather, in the tree's vast shade,
To honor the legacy that nature made.
For the last standing tree, in its lonely stand,
Is a testament to life, and the beauty of land.

Let's plant new seeds, let's water with care,
Let's grow a new forest, fresh and fair.
For in the last standing tree, there's a dream to weave,
Of a world reborn, for us to achieve.

This poem awakens a deep reverence for the natural world and fosters a commitment to reforestation and environmental stewardship, urging us to nurture and protect our planet with heartfelt dedication.

68. The Melting Mountains

The Melting Mountains, once proud and tall,
Now weep as they heed the warm call.
Their icy crowns, once gleaming white,
Shed tears of blue in the sun's harsh light.

Majestic peaks of silent stone,
Now echo with a somber tone.
The Melting Mountains, a cry so stark,
A beacon of change, a vanishing mark.

Glaciers retreat, rivers swell,
Stories of old, the mountains tell.
Of a time when snow would softly rest,
Upon their shoulders, a wintry vest.

But warmth encroaches, day by day,
The Melting Mountains' strength gives way.
A cascade of change, a flowing plight,
In the mountain's tears, a reflected light.

Let us hear their silent plea,
In the trickle of ice, a call to see.
The Melting Mountains need our care,
For their frozen song fills the air.

So let us act, let us rise,
To save the mountains, the cold, the ice.
For in their beauty, a world so bright,
The Melting Mountains, our guiding light.

This poem stands as a poignant reminder of the fragile majesty of our planet's mountains and the urgent need to confront the impacts of climate change, inspiring us to act with resolve to protect these towering guardians of our world.

69. The Peaceful Peaks

Atop the cradle where the daybreak wakes,
Stands the tranquil guardian, Mt. Kenya takes.
Its peak so noble, draped in white,
A silent sentinel in the early light.

The air is crisp, the sky so clear,
Above the clouds, the peak draws near.
A haven for life, both lush and rare,
In the peaceful peak's protective care.

The elephant roams, the leopard strides,
In the shadow of the mountain's sides.
The flora blooms, a vibrant sight,
Nurtured by the peak's gentle might.

But even here, where peace seems to reign,
The whispers of change are not in vain.
The glaciers weep, the snowcaps thin,
A warming world, now closing in.

Let us honor the peak, so calm and bold,
By protecting its story, yet to be told.
For Mt. Kenya's peace is ours to keep,
A pledge of respect, a promise deep.

This poem captures the serene majesty of Mt. Kenya, celebrating its grandeur and underscoring the crucial importance of preserving such natural wonders for the benefit of future generations.

70. The Placid Plains

The Placid Plains, a canvas wide,
Where horizons stretch, and hearts abide.
A tranquil sea of grass and grain,
The Plains' soft song, a soothing refrain.

Gentle hills, a subtle rise,
Under the vast, unending skies.
The Placid Plains, in whispers low,
Tell tales of peace, where soft winds blow.

Here, the earth breathes, slow and deep,
In the Plains' embrace, the world does sleep.
A harmony of land, of sky, of air,
In this quiet place, life is fair.

The dance of light, on fields so still,
In the Placid Plains, time takes its fill.
A symphony of silence, a calm so rare,
In the Plains' grace, no burden to bear.

Yet, the Plains plead, in their silent way,
For respect, for care, for a brighter day.
For the Placid Plains, so vast, so wide,
Hold the dreams of earth, inside.

So let us honor, let us keep,
The Placid Plains, in their slumber deep.
For in their quiet, in their serene,
Lies the heart of the world, so pure, so clean.

This poem reflects the gentle beauty and profound tranquility of the Placid Plains, inspiring a deep sense of stewardship and a commitment to preserving these vast and peaceful landscapes for future generations.

71. The Quiet Quarry

The Quiet Quarry, a scar so still,
Where once was noise, now hush does fill.
A place of toil, of stone, of might,
Now rests in silence, day and night.

Walls of rock, carved deep and wide,
In the Quiet Quarry, shadows hide.
Echoes of the past, whispers of the drill,
In its stillness, time seems to still.

Water fills the basin, clear and deep,
In the quarry's heart, secrets to keep.
A new life stirs, in this quiet berth,
Nature reclaims, asserts its worth.

The Quiet Quarry, a testament to time,
To the earth's resilience, to the climb.
From the depths, a lesson drawn,
In the quietude, a new dawn.

Let us learn from the quarry's hush,
In its tranquil depths, the world's lush.
For in its silence, a story told,
Of a future bright, of a vision bold.

This poem reflects the transformation and tranquility of a quarry reclaimed by nature, symbolizing resilience and hope amidst change, and inspiring us to find beauty and renewal in the cycles of our environment.

72. The Radiant Reef

The Radiant Reef, a jewel beneath the waves,
A mosaic of life, where the ocean saves.
Corals bloom in colors so bright,
In the Reef's soft glow, the sea's delight.

Fish dart through the water's hue,
A shimmering dance, a vibrant crew.
Anemones sway, starfish rest,
In the Radiant Reef, the ocean's best.

A symphony of shapes, of textures so rare,
In the Reef's embrace, there's no despair.
A sanctuary for the small, the sleek,
In its radiant halls, they play hide and seek.

Yet, the Reef whispers of a fragile fate,
Of warming seas, of a state so dire.
The Radiant Reef, in its silent plea,
Calls for care, for a future to see.

Let us heed the Reef's hushed call,
To protect, to preserve, for one and all.
For in its radiance, a world so grand,
The Reef's beauty, our sea's land.

This poem celebrates the vibrant beauty and ecological significance of coral reefs, highlighting the urgent need to protect these underwater wonders. It calls us to safeguard the rich diversity of life they sustain, inspiring a commitment to preserving these fragile marvels of the ocean.

73. The Rising Ridge

Upon the earth, where horizons merge,
Stands proud and tall, The Rising Ridge.
A backbone of stone, reaching for the skies,
A monument of nature, where the eagle flies.

Its slopes, a patchwork of shrub and tree,
A bastion for life, wild and free.
The ridge ascends, with stories untold,
Of ancient times and warriors bold.

The wind sings high on the rocky crest,
A lullaby for the land's verdant nest.
The sun casts shadows, long and wide,
Over the ridge, nature's own pride.

But now the ridge feels the warming sun,
A sign of battles, yet to be won.
The climate shifts, the ridge stands witness,
To the need for change, and human fitness.

Let's honor the ridge, its rise and fall,
By guarding the climate, for one and for all.
For The Rising Ridge, in its lofty flight,
Reminds us to keep our future bright.

This poem inspires a profound respect and sense of responsibility towards our planet's magnificent landscapes, emphasizing the urgent need to confront environmental challenges and protect these awe-inspiring treasures for future generations.

74. The Roaring River

Where mountains part and valleys bend,
The roaring river, nature's friend,
A force of life, a pulse so strong,
In its mighty course, we belong.

It carves the earth with liquid might,
A sculptor's hand in the light,
From humble springs to the ocean's call,
The roaring river gives its all.

Its voice is heard in rapids wild,
A timeless song, nature's child,
Where salmon leap and otters play,
The river roars in its own ballet.

But this roar is not just a sound to hear,
It's a call to protect what we hold dear,
For as the river thunders on its way,
It speaks of choices we make each day.

So let us honor this watery roar,
For it tells of health, of life, and more,
May the roaring river forever flow,
A testament to the life we know.

This poem fosters a deep respect for the dynamic, life-sustaining rivers that grace our planet, urging us to honor and protect these vital waterways that nurture and sustain the tapestry of life.

75. The Rustling Reeds

The Rustling Papyrus Reeds, a whispering throng,
In the dance of the breeze, they sway and they throng.
A symphony of rustles, a hushed, paper song,
In the heart of the wetlands, where they belong.

Tall and slender, in clusters they stand,
The Papyrus Reeds, a marshland band.
Their secrets inscribed on ancient scroll,
In the rustling leaves, their stories unroll.

By the water's edge, in the mud and the mire,
The Reeds grow strong, never to tire.
A habitat for bird, for fish, for frog,
In the Rustling Papyrus, a natural log.

Yet, they speak of times, of waters receding,
Of the earth's quiet plea, its gentle pleading.
For care, for thought, for a hand to sow,
The Rustling Papyrus Reeds, in the wind's low blow.

So let us listen to the Reeds' soft speech,
In their rustle, a lesson they teach.
To preserve, to protect, to let live,
The Rustling Papyrus Reeds, their gift to give.

This poem captures the delicate beauty and ecological significance of papyrus reeds, swaying with timeless wisdom. It reminds us to heed the subtle yet profound voices of nature and cherish the ancient wisdom they impart.

76. The Sacred Sands

In the heart of the desert, under the sun's fierce gaze,
Lie the Sacred Sands, where the ancient spirits blaze.
A sea of dunes, shifting with the wind's command,
A timeless realm, a silent, hallowed land.

Golden grains tell tales of old,
Of caravans, bold and untold.
A canvas vast, nature's own hand,
Sculpting art in the Sacred Sands.

Here, the cactus stands resilient and strong,
And the night's sky sings the stars' song.
Life thrives in this arid expanse,
A testament to nature's intricate dance.

But these sands whisper a foreboding truth,
Of encroaching change, of lost youth.
The dunes may shift, but should not fade,
Under the weight of human-made shade.

Let us tread lightly on this sacred ground,
Where every grain of sand is profound.
For in its silence, wisdom resides,
In the Sacred Sands, where eternity abides.

This poem evokes the profound beauty and spiritual significance of the desert sands, highlighting the importance of respecting and preserving these delicate ecosystems. It inspires us to honor the unique wonders of the desert and commit to safeguarding its fragile balance.

77. The Sacred Skies

Above the earth, in the canvas vast,
The Sacred Skies hold the future and past.
A dome of blue, with clouds that glide,
Carrying dreams on a celestial tide.

The sun's rays pierce with golden light,
While stars twinkle back the tales of night.
A breath of the divine, in the air so high,
In the Sacred Skies, where hopes can fly.

Eagles soar on the breeze, so free,
In the endless expanse, where souls agree.
To respect the air, the wind, the flight,
In the Sacred Skies, our shared delight.

But these skies, they echo a silent plea,
From the whisper of wind to the buzz of bee.
For purity lost, for clarity dimmed,
By the smog of progress, the horizon skimmed.

Let's gaze above, with reverence anew,
And pledge to the skies, a vow so true.
To guard the sacred, the azure dome,
In the Sacred Skies, our planet's home.

This poem inspires a deep sense of awe and responsibility towards the vast, beautiful skies that embrace our world, encouraging us to cherish and protect the boundless expanse that sustains and inspires us.

78. The Secret Stream

Whispering waters of The Secret Stream,
Flowing unseen, like a quiet dream.
Through the hush of the forest's embrace,
It carves a path with gentle grace.

The Secret Stream, a hidden gem,
In nature's diadem, a rare stem.
Its banks are lush with ferns and moss,
A world away from human loss.

Beneath the canopy, it winds and waves,
A tapestry that Mother Earth conceives.
With every babble, ripple, and gleam,
It tells a tale, The Secret Stream.

Its crystal waters, pure and clear,
A sanctuary for life to revere.
Fish dart below, in playful chase,
In this serene, secluded place.

The Secret Stream, with its soft caress,
Heals the land with its tenderness.
A nurturing force, so calm and serene,
Bestowing life, yet remains unseen.

So let us guard this stream with care,
For its secrets are ours to share.
A testament to the unseen,
The vital beauty of The Secret Stream.

This poem flows with the same tranquil grace and hidden beauty as The Secret Stream, mirroring its serene charm and inspiring us to seek out and cherish the quiet wonders that flow through our world.

79. The Serene Savanna

The Serene Savanna, a breath so wide,
Where horizons stretch, and dreams reside.
A tapestry woven with gold and green,
In the Savanna's calm, life's serene.

Acacias dot the sprawling land,
Under the sky, they proudly stand.
The Serene Savanna, whispers of old,
Of a world untouched, of stories untold.

Giraffes stride in elegant grace,
Each step a gentle, loping embrace.
Zebras graze in the golden light,
In the Savanna's peace, there's no fright.

The lion's roar, a distant sound,
In the Serene Savanna, peace is found.
A harmony of life, under the sun's dome,
In this tranquil land, many call home.

Yet, the Savanna speaks, in a voice so mild,
Of the need for care, for the wild.
For the Serene Savanna, vast and free,
Is a treasure to guard, for you and me.

So let us honor this quiet plain,
Where the earth sings in a soft refrain.
For in the Serene Savanna's gentle hand,
Lies the heart of the wild, the soul of the land.

This poem captures the tranquil beauty and delicate balance of life in the Serene Savanna, reminding us of the vital importance of preserving these open spaces to support and celebrate the rich diversity of life they nurture.

80. The Silent Indian Ocean

Beneath the sky, so vast and blue,
Lies the Indian Ocean, silent and true.
Its waves once roared with mighty glee,
Now whisper tales of what used to be.

Coral reefs, once vibrant, bright,
Now fade away in the dimming light.
The fish that swam in schools so grand,
Are dwindling now, by human hand.

The monsoon winds that danced and played,
Now carry whispers of dismay.
Plastic tides and oil spills,
Mar the beauty of its hills.

Yet in the silence, hope remains,
For nature's strength, it still sustains.
If we listen to the ocean's plea,
We might restore its majesty.

The turtles, dolphins, whales, and more,
Await the day their home will soar.
With cleaner waters, skies so clear,
The Silent Indian Ocean will persevere.

So let us heed the silent call,
To protect, preserve, and cherish all.
For in our hands, the power lies,
To save the ocean, beneath the skies.

This poem evokes a profound reverence for the Indian Ocean's serene beauty, inspiring a deep commitment to preserving its vital ecosystem and ensuring its enduring health for future generations.

81. The Silent Sky

Above the world, so high and wide,
The silent sky, in its tranquil pride.
A canvas of blue, where dreams take flight,
A quiet guardian, through day and night.

No words are spoken, yet it tells a tale,
Of the sun's warm kiss, the moon's white veil.
The silent sky, a witness to all,
The rise and fall, the big and small.

But look closer, see the stars grow dim,
As the lights below, make the heavens grim.
The silent sky, once clear and bright,
Now veils its face, from the city's light.

The hush of the sky, is a plea, a sign,
To remember the stars, let them shine.
For the silent sky, holds the key,
To a universe vast, a mystery.

Let's turn down the lights, let darkness fall,
And listen to the sky, to its silent call.
For in its quiet, there's a song to hear,
An environmental hymn, crystal clear.

So let us cherish, the silent sky,
And work to protect, the highs and the lows.
For the sky's silence, is a precious gift,
A reminder of the peace, we're here to lift.

This poem inspires a sense of stewardship for the night sky and the importance of reducing light pollution to preserve the natural wonder of the stars.

82. The Silent Springs

In the hush of dawn, where waters flow,
The Silent Springs, in whispers low.
A gentle murmur, a tranquil tune,
Played on the harp of the crescent moon.

These springs, they speak in ripples soft,
Of secrets kept aloft.
In their quiet, life abounds,
In every splash, nature's sounds.

The deer sip clear, the lilies bloom,
Around the springs, life resumes.
A ballet of fish, a chorus of frogs,
In this watery haven, away from the fogs.

But silence grows, as shadows fall,
The springs, they heed a different call.
A call to arms, against the tide,
Of silent threats, far and wide.

Let's guard the springs, their silent song,
For in their purity, we belong.
In the quiet of the springs, we find,
A reflection of our own peace of mind.

This poem captures the serene essence of springs, highlighting the profound importance of safeguarding these delicate ecosystems. It reminds us of the tranquility they bring to both nature and our souls, inspiring a deep commitment to their preservation.

83. The Silent Stream

In the quiet woods, where shadows play,
A silent stream winds its gentle way.
A murmur soft, a liquid gleam,
A hidden jewel, a whispered dream.

Its waters clear as the purest glass,
Over pebbles and stones, it softly passes,
A haven for life in its tender stream,
Where fish glide silent as a daydream.

The banks are lush with mosses green,
Where footprints are rare, and air is clean,
The stream, it speaks without a word,
A tale of peace, too seldom heard.

It quenches the thirst of deer at dawn,
Reflects the sky as day is born,
It's where the heron finds its meal,
A place where time seems to heal.

But silent streams are not just water,
They're life's essence, nature's daughter.
They carry stories from the mountain's heart,
Of snow and rain, where rivers start.

So let's protect these streams so still,
From harm and hurt, from ill and spill,
For in their silence lies a song,
Of a world where all can belong.

May the silent stream forever flow,
A testament to the less we know,
A reminder to tread with care,
For in its silence, life is there.

This poem beautifully captures the tranquil yet essential presence of a stream within our environment, celebrating its quiet strength and the vital role it plays in sustaining the world around us.

84. The Singing Sand

In the heart of the desert, under the sun's command,
Rises a melody known as the singing sand,
A whisper of grains, a desert chant,
An environmental tale, enchanting and scant.

Each grain a note in the dune's soft song,
Together they hum, where they belong,
A symphony played by the wind's soft hand,
In the vast, open stage of the singing sand.

The dance of particles, in the heat's embrace,
Tells a story of survival, of time and space,
Where water is scarce, and life is rare,
In the singing sand, beauty is bare.

But this song is more than a desert's tune,
It's a call to respect, to act, and soon,
For as we listen to the sand's soft sigh,
We must protect these dunes, under the sky.

So let us treasure this sandy lore,
For in its song, there's wisdom and more,
The singing sand, with its voice so grand,
Reminds us to care for this fragile land.

This poem evokes the unique beauty and haunting song of the desert sands, reminding us of the delicate balance we must strive to maintain with our environment, and inspiring us to honor and protect these fragile landscapes.

85. The Singing Summit

Atop the world, where eagles dare to fly,
A peak stands proud, beneath the vast sky.
The Singing Summit, a name it's fondly given,
For whispers of wind, through crags are driven.

Majestic and tall, cloaked in emerald grace,
A guardian of time, in Earth's embrace.
Its slopes, a canvas of nature's grandeur,
Home to the wild, pure and demure.

From its crest, the trees sway and dance,
To the rhythm of life, in a verdant trance.
The summit sings, with a voice so clear,
A song of the wild, for those who hear.

It tells a tale of the ages past,
Of ice and fire, in shadows cast.
A witness to change, yet firmly it stands,
A testament to time's unyielding hands.

But hark, the summit's song grows faint,
As scars of progress, paint its quaint.
The cry for help, in each note that falls,
A plea to protect, its hallowed halls.

So let us heed, The Singing Summit's call,
To cherish and guard, for one and for all.
For in its tune, the truth is spun,
We're not apart, but a part of one.

This poem resonates with the beauty of our natural wonders and underscores the profound importance of preserving them, inspiring us to cherish and protect these precious gifts for generations to come.

86. The Throbbing Thicket

In the heart of the land, where wild winds whistle,
Lies the Throbbing Thicket, nature's own bristle.
A dense, lush haven, where life's choir sings,
A symphony of green, on delicate wings.

The Throbbing Thicket, pulsing with life,
A bastion against the world's strife.
Roots gripping earth, branches entwined,
A fortress of flora, intricately designed.

Leaves whisper secrets, in rustling tones,
Over mossy stones, where the clear stream moans.
A dance of shadows and light, beneath the sun's flicker,
In the Throbbing Thicket, time moves quicker.

Here, beetles scuttle and birds take flight,
In a dazzling display from morning till night.
The air is alive with the buzz of the bees,
And the sweet scent of flowers, carried on the breeze.

The Throbbing Thicket, a world apart,
Beats with the rhythm of the Earth's heart.
A reminder to all, of the beauty we're given,
In this verdant space, where life is truly livin'.

So let's cherish this thicket, with all its might,
For it holds the key to our planet's plight.
A green jewel in our hands, to protect and respect,
The Throbbing Thicket, our environment to reflect.

This poem enriches the poetic essence of your work, elevating its beauty and depth to inspire a deeper appreciation and connection with the natural world.

87. The Thundering Thaw

When winter's grip begins to wane,
And spring whispers life's refrain,
The Thundering Thaw with might unfurls,
In cascades swift, the water swirls.

From mountain high to valley deep,
The snow's retreat makes rivers weep.
A symphony of cracks and roars,
Nature's force on open doors.

The ice that clung with frosty claws,
Now yields to time's unyielding laws.
The thaw, it thunders, shakes the ground,
A testament to cycles found.

Beneath the white, a promise green,
A world reborn, a sight unseen.
Life stirs beneath the melting fray,
As night gives way to longer day.

Yet, in this thaw, a warning tolls,
For warming trends and shifting poles.
The thundering may speak of dread,
Of changing climes and futures read.

So let us listen, learn, and act,
To balance back the thaw's impact.
For in the thunder's powerful draw,
Lies the Earth's plea—a call to awe.

This poem reflects the powerful transformation of the thawing season, revealing the broader environmental implications of our changing climate. It inspires us to recognize and act upon the profound shifts occurring in our world, urging us to embrace and address these challenges with urgency and hope.

88. The Tranquil Tropics

In the realm where the Tranquil Tropics lie,
Underneath the ever-stretching sky.
The air hums with a vibrant tune,
In the land where it's always June.

Palms sway with a graceful ease,
Whispering the secrets of the sultry breeze.
The sun kisses the earth with a golden glow,
In the Tropics, where life flows slow.

The canopy teems with chattering life,
A world away from human strife.
Birds of paradise, in colors bold,
In the Tranquil Tropics, stories unfold.

Fruits hang heavy, ripe and sweet,
A succulent treasure, a jungle's treat.
The air is thick with the scent of bloom,
In the Tropics' embrace, there's no room for gloom.

Yet, the Tropics speak in a silent plea,
For respect, for care, for empathy.
For the hand of man can take or give,
In the Tranquil Tropics, let all life live.

So let's preserve this haven of peace,
Where the whispers of nature never cease.
For in the Tranquil Tropics' gentle shade,
Is a paradise that must never fade.

This poem encapsulates the serene beauty and delicate balance of tropical regions, urging us to preserve these lush sanctuaries that sustain a vibrant tapestry of life. It inspires a deep commitment to protecting these vital havens for the countless species that call them home.

89. The Veiled Valley

The Veiled Valley, shrouded in mist,
A realm of wonder, by the sun's light kissed.
A cloak of fog, a soft embrace,
In the valley's veil, a hidden grace.

Whispers of water, the rustle of leaves,
In the Veiled Valley, the earth breathes.
A tapestry of green, a silent bower,
Where time stands still, hour by hour.

Shadows play on the valley's floor,
A dance of light, nature's lore.
The Veiled Valley, a secret kept,
In its gentle folds, the dew has wept.

Yet, beneath the veil, a world thrives,
A bustling hub of countless lives.
From the tiniest ant to the soaring hawk,
In the valley's shroud, life does walk.

The Veiled Valley, a lesson to learn,
Of the beauty that lies at every turn.
For in its quiet, in its calm,
The valley sings a soothing psalm.

So let us cherish this veiled land,
With a gentle heart, with a caring hand.
For the Veiled Valley, so lush, so wild,
Is nature's poem, softly compiled.

This poem captures the serene and mystical essence of the Veiled Valley, revealing the hidden beauties that dwell within the natural world. It inspires us to seek out, appreciate, and protect these enchanting wonders that await our discovery and care.

90. The Verdant Valley

Nestled 'neath the watchful peaks,
Lies a valley, where the earth speaks,
In shades of green, a vibrant quilt,
Where life is treasured, and time is spilt.

The Verdant Valley, lush and wide,
A cradle of nature, in pride it bides,
With rivers meandering, a silken thread,
Through meadows rich, where deer tread.

Trees arch high, a leafy dome,
In this verdant valley, creatures roam,
Birds trill songs from dawn till dusk,
In this slice of paradise, pure and brusque.

Flowers bloom in a riot of hues,
A palette of colors, morning dews,
The air is sweet with the scent of bloom,
In the valley's embrace, there's no gloom.

But this verdant haven, so serene,
Faces threats unseen, unclean,
We must guard this valley's grace,
For in its health, our own we trace.

So let us sing the valley's song,
A melody where we all belong,
In harmony with the land so fair,
The Verdant Valley, our solace, our care.

This poem fosters a deep connection with the natural beauty of valleys, highlighting their vital role in nurturing diverse ecosystems. It inspires us to cherish and protect these magnificent landscapes that sustain a rich tapestry of life.

91. The Wailing Waves

The Wailing Waves, with their mournful cry,
Speak of the tales that we can't deny.
A lament for the deep, a siren's song,
Where the ocean's heart has been strong for long.

The Wailing Waves, in their restless churn,
Echo the concerns for which we yearn.
A plea to the shore, to the sky, to the sands,
To heed the message that the sea demands.

With each crash and retreat, a story told,
Of human deeds, both brash and bold.
The waves wail for the coral bleached white,
For the marine life that fades from sight.

Yet, in their sorrow, there's a call to action,
A rallying force, a chain reaction.
To guard the seas, the creatures' haven,
From the storm of harm, we must awaken.

The Wailing Waves, a symphony of care,
A reminder of the duty we all share.
To preserve the blue, the life below,
For the waves to sing, not wail, in their flow.

So let us listen to the ocean's plea,
In the wailing waves, a lesson to see.
To cherish the water, the tide, the wave,
For a future of blue, beautiful and brave.

This poem captures the ocean's heartfelt plea, inspiring a steadfast commitment to environmental stewardship and urging us to protect and preserve the vast, vital waters that sustain our planet.

92. The Waterfall's Whisper

In the heart of the wild, where the waters cascade,
The waterfall's whisper, in light and shade.
A symphony of droplets, in a ceaseless pour,
Telling tales of the earth, of myth and lore.

The whisper is a song, of power and might,
A voice of the ages, through day and night.
It speaks of the mountains, the journey so steep,
Of the rivers and streams, the canyons deep.

The mist that rises, a breath so fine,
Carries the whisper, through the pine.
A message of purity, of nature's choir,
The waterfall's whisper, never to tire.

But this whisper also holds, a warning clear,
Of the waters at risk, that we hold dear.
A call to protect, to preserve and to heed,
The waterfall's whisper, in its time of need.

So let us listen, with hearts open wide,
To the waterfall's whisper, a guide and a tide.
For in its voice, lies the essence of life,
A call to action, in the world's strife.

This poem kindles a profound sense of awe for the majestic beauty of waterfalls, highlighting the vital importance of conserving the precious water resources they symbolize. It calls us to honor and protect these natural wonders that embody both splendor and life-sustaining power.

93. The Weeping Wetlands

The Weeping Wetlands, where teardrops fall,
A realm of water, a living shawl.
Draped over the earth, in layers so deep,
Where the echoes of nature softly weep.

In the Weeping Wetlands, the reeds do sigh,
Under the watch of the weeping sky.
A cradle of life, in the marshy hold,
A story of survival, ages old.

Mists rise and fall, in a ghostly dance,
In the wetlands' expanse, a silent trance.
The croak of a frog, the splash of a fin,
In this watery world, life begins.

The Wetlands weep for the loss they bear,
For the scars of the land, laid bare.
A plea for mercy, for care, for thought,
For the battles fought, and the lessons taught.

Yet, in their sorrow, there's hope that gleams,
In the Weeping Wetlands, life teems.
A sanctuary for wing and webbed feet,
In the dance of the rain, the cycle's complete.

So let us heed the Wetlands' cry,
For in their tears, the truth does lie.
To protect and preserve, to honor and save,
The Weeping Wetlands, so brave.

This poem serves as a gentle reminder of the delicate balance within wetlands and the crucial importance of preserving these vital ecosystems. It inspires us to safeguard these rich habitats, which support a diverse array of life, ensuring their resilience and beauty for future generations.

94. The Whispering Wild

The Whispering Wild, a chorus unseen,
A breath of the earth, pure and serene.
Where the trees converse in rustling leaves,
And the secrets of nature, the wild weaves.

In the hush of the dawn, the wild's soft call,
A delicate murmur, a natural thrall.
The rustle of grass, the creek's gentle prattle,
In the Whispering Wild, there's no need for battle.

The wild speaks in tones so low and profound,
In the flutter of wings, in the soft ground.
A whisper of life, of cycles that spin,
In the wild's embrace, where wonders begin.

Creatures roam with a tread so light,
Through the Whispering Wild, from day to night.
In their silent language, they speak of care,
Of a world that's precious, rare, and fair.

The Whispering Wild, with its subtle power,
Reminds us to cherish every tree, every flower.
For in its quiet strength, there's wisdom to find,
In the gentle whispers of the wild, so kind.

So let us listen, with hearts wide open,
To the Whispering Wild, and its quiet token.
For in its hushed tones, lies a message clear,
Protect the wild, hold it dear.

This poem ignites a deeper appreciation for the subtle yet profound voice of the wild, urging us to protect the natural world that quietly imparts its wisdom. It calls us to listen and act with reverence, safeguarding the whispers of nature that guide and inspire us.

95. Whisper of the Trees

In the forest's heart, where shadows play,
The trees whisper softly, come what may.
A rustling language, old as time,
A verdant chorus, a natural rhyme.

Leaves speak in hushes, to the sky so wide,
Telling tales of the earth, in green pride.
The whisper of the trees, a secret lore,
Of life's deep roots, and so much more.

They murmur of rain, of sun's warm kiss,
Of seasons changing, of nature's bliss.
The whisper of the trees, in gentle sway,
A reminder to cherish, come what may.

But these whispers now, bear a somber tone,
Of the scars we've left, of the seeds we've sown.
They speak of loss, of branches bare,
A plea for healing, for tender care.

So let's heed the whisper, of the leafy boughs,
To protect the trees, to make our vows.
For the whisper of the trees, in its subtle might,
Is the voice of the earth, of the day and night.

This poem captures the essence of the trees' silent yet profound communication, emphasizing the importance of attuning ourselves to their whispers. It inspires us to listen with reverence and commit to preserving these ancient voices for the well-being of future generations.

96. Whispering Willows

In the hush of dawn's first light,
Where willows stand, a serene sight,
Their branches sway with a grace so still,
In the quiet air, a tranquil thrill.

Leaves that whisper secrets old,
In silvery green, stories told,
Of riverside and meadow's edge,
Where willows guard the water's pledge.

Their roots, a network deep and wide,
In the earth's embrace, they confide,
A life source for the soil and stream,
In the whispering willows, life does gleam.

The willow's song is soft and low,
A melody only the heart can know,
It speaks of balance, of ebb and flow,
In the language of the wind's soft blow.

But these gentle giants, tall and fair,
Face a world that does not care,
We must listen to their leafy prose,
And protect the willows, where the river flows.

So let us cherish these trees so fine,
For in their whispers, a truth divine,
The willows teach us to bend, not break,
To live in harmony, for the planet's sake.

This poem evokes a profound sense of peace and deepens our appreciation for willow trees, celebrating their graceful presence and the vital role they play in our environment. It inspires us to cherish these natural guardians and honor their contribution to the beauty and balance of our world.

97. Whispers in the Woods

In the woods where shadows play,
And light dapples through the day,
The whispers rise from leaf and limb,
A woodland chorus, a natural hymn.

The trees, they speak in rustling tones,
Of sunlit clearings and mossy stones,
Their leaves tell tales of rain and sun,
Of seasons changing, of life begun.

The woods are alive with secret words,
In the flight of birds, the scurry of herds,
Each whisper a thread in the fabric of life,
A tapestry woven without strife.

But these whispers carry a deeper sound,
A plea for the woods, for the sacred ground,
For as we tread through the forest deep,
It's the woods' secrets we must keep.

So let us listen with heart and soul,
To the whispers in the woods, to the stories told,
And pledge to protect this haven so rare,
For the whispers in the woods, are a prayer.

This poem resonates with the mystical and essential presence of the woods in our environment, capturing their enchanting allure and the crucial role they play in sustaining the natural world. It inspires us to revere and protect these forested sanctuaries, which hold the magic and vitality of our planet.

98. Whispers of the Plains

Across the vast and open land,
Where golden grasses gently stand,
The whispers of the plains arise,
Beneath the wide and endless skies.

Once teeming with life, so wild and free,
Now echoes of a memory.
The buffalo that roamed in herds,
Are now but shadows, fleeting birds.

The wind that sweeps the rolling hills,
Carries tales of ancient thrills.
Of prairies lush and rivers clear,
Now fading fast, year by year.

The flowers bloom in muted hues,
As if they mourn the morning dews.
The soil, once rich, now dry and bare,
Cries out for tender, loving care.

Yet in the silence, hope persists,
In every breeze, in every mist.
For nature's heart still beats within,
The plains await their life to begin.

If we can hear the whispers' plea,
To heal, to nurture, to set free,
The plains may once again arise,
Beneath the wide and endless skies.

So let us heed the call of earth,
To cherish, guard, and prove its worth.
For in our hands, the future lies,
To save the plains, beneath the skies.

This poem evokes the serene and timeless spirit of the plains, illuminating the urgent need to listen to and safeguard these vast natural spaces. It inspires us to honor their quiet beauty and preserve their enduring essence for future generations.

99. Whispers of the Wind

Through valleys deep and mountains high,
The wind it whispers, a gentle sigh.
It carries tales of earth and sky,
Of days gone by, of reasons why.

Once it danced through forests green,
Where every leaf and branch was seen.
Now it mourns the trees that fell,
To human hands, a silent knell.

It sweeps across the desert sands,
Where life once thrived in vibrant bands.
Now dunes shift in endless plight,
A testament to nature's fight.

The oceans hear its mournful song,
Of coral reefs where life belonged.
Now bleached and broken, they remain,
A silent cry, a world in pain.

Yet in the whispers, hope is found,
In every gust, in every sound.
For nature's voice, though soft and low,
Speaks of a future we can sow.

If we can hear the wind's soft plea,
To heal, to mend, to set things free,
The earth may once again renew,
Beneath the skies of endless blue.

So let us heed the whispers' call,
To cherish, guard, and save it all.
For in our hands, the power lies,
To heal the earth, beneath the skies.

100. Whispers of Wilderness

Whispers of Wilderness, soft and serene,
A chorus of life, unheard, unseen.
In the depth of the woods, where shadows play,
Nature's secrets are whispered away.

The Wilderness speaks in a language so old,
In the rustling leaves, in the mossy mold.
A hush of wonder, a breath of air,
In the Wilderness, whispers are everywhere.

Trees converse in creaks and sighs,
While the brook murmurs as it flows by.
The call of the wild, so faint, so nigh,
In the Whispers of Wilderness, under the sky.

Creatures tread with silent paws,
Abiding by nature's unwritten laws.
In their eyes, a world untold,
In the Wilderness, stories unfold.

The Whispers of Wilderness, a sacred song,
A plea for respect, to right the wrong.
For in its quiet, lies strength untold,
In the whispers, wisdom of ages old.

So let us listen, let us heed,
To the Wilderness's whisper, in every deed.
For in its voice, so soft, so mild,
Lies the future of every child.

This poem echoes the subtle yet profound voice of the wilderness, urging us to listen attentively and safeguard the natural world that speaks to us in gentle whispers. It inspires us to honor and protect the delicate messages of nature that guide and enrich our lives.

WHISPERS OF NATURE: 100 POEMS ON CLIMATE AND ENVIRONMENT
Whispers of Earth: Poetry Unveiled.
Journey through sunsets, mountains, and wildlife encounters. Each verse is a tapestry of our planet's beauty and the urgency of preservation. Seek solace, inspiration, and connection as you traverse meadows, cliffs, and star-studded skies. 🌿 🌍 🦋

SAVVY SAVANNA ANIMALS IN BUSINESS
Unlock Business Success: Learn from the Wild Wisdom of Animals!
The author unveils invaluable lessons from the animal kingdom. Discover adaptability through diverse environments and the strength of bonds exemplified by wolves. Whether you're a startup founder or seasoned executive, explore practical insights inspired by nature's stealthy falcons and bustling ant colonies. ✺📚🦉

SEEN AND THE UNSEEN
Unveil the Hidden. Embrace the Seen:
Discover purpose, identity, and cosmic connections.

This journey bridges spiritual realms and earthly existence, weaving contemporary challenges with ancestral wisdom. Dive into growth, resilience, and empathy as unseen forms evolve into tangible reality. ✺✺

101. Contact the Author

Phone:

- (+254) 0722 711684
- (+254) 0732 341324

Email: victor254isaacs@gmail.com

Social Media:

- ❖ **Twitter**   : Victor Isaacs254
- ❖ **Instagram** : Victor Vsaacs254
- ❖ **Facebook**   : Victor Isaacs254
- ❖ **YouTube**   : Victor isaacs254
- ❖ **LinkedIn**   : Victor Isaacs
- ❖ **Website**   : www.alihsum.com

www.ingramcontent.com/pod-product-compliance
Lightning Source LLC
Chambersburg PA
CBHW050318230526
45471CB00005B/2246